THE LIVES OF

LICHENS

THE LIVES OF LICHENS

A NATURAL HISTORY

Robert Lücking & Toby Spribille

PRINCETON UNIVERSITY PRESS
PRINCETON AND OXFORD

Published in 2024 by Princeton University Press
41 William Street, Princeton, New Jersey 08540
99 Banbury Road, Oxford OX2 6JX
press.princeton.edu

Copyright © 2024 by UniPress Books Limited
www.unipressbooks.com

Library of Congress Control Number 2023943722
ISBN 978-0-691-24727-4
Ebook 978-0-691-24728-1

Typeset in Bembo and Futura

Printed and bound in Malaysia

10 9 8 7 6 5 4 3 2 1

British Library Cataloging-in-Publication Data is available

This book was conceived, designed, and produced by
UniPress Books Limited
Publisher: Nigel Browning
Commissioning editor: Kate Shanahan
Project manager: David Price-Goodfellow
Designer & art direction: Wayne Blades
Copy editor: Hugh Brazier
Picture researcher: Elaine Willis
Illustrators: John Woodcock and Chelsea Lau
Maps: Les Hunt

Cover images: Photographs by Dr. Robert Lücking
Previous page, main image: *Ramalina menziesii*

CONTENTS

The world of lichens

For thousands of years, humans have looked at lichens, touched them, sniffed them, tasted them, ground them into powder for tea, dyed textiles with them, distinguished different kinds, observed where they grow and where they don't. We cannot know for sure what our ancestors thought of them, but surely they have always caused people to think. Are they plants? Seaweeds, adapted for life on land? Strange forms of fungi?

We now know that lichens are a symbiosis, two or more organisms in one. The bulk of a lichen is a fungus, the mycobiont, embedded within the fabric of which are tiny photosynthesizing cells of algae and/or cyanobacteria, the photobionts. A whole field,

lichenology, has sprouted up around their study, and a lexicon has been created for their symbionts and their endless forms and structures.

Getting to our present understanding of lichens, such as it is, has been a long road. The study of lichens is one of patience and attention to detail, of nature walks that slow to crawls, of magnifying lenses held to the eye. As we have taken them apart and asked what they are made of, how they can be so resilient and yet so sensitive, how they can be both many and one, lichens have themselves served as looking glasses through which we re-envision our own worlds.

In this book, we seek to provide an entry point for those who wish to learn more about these fascinating beings and endeavor to enter their alien worlds. Lichens clothe nearly all of the terrestrial Earth, and we cannot possibly tell all their stories or illustrate all their marvelous forms. But we hope that we can here provide some basic information about their biology, their ecology and diversity, their study, and their importance to nature and to us humans, to help the reader gain a greater appreciation of the lives of lichens.

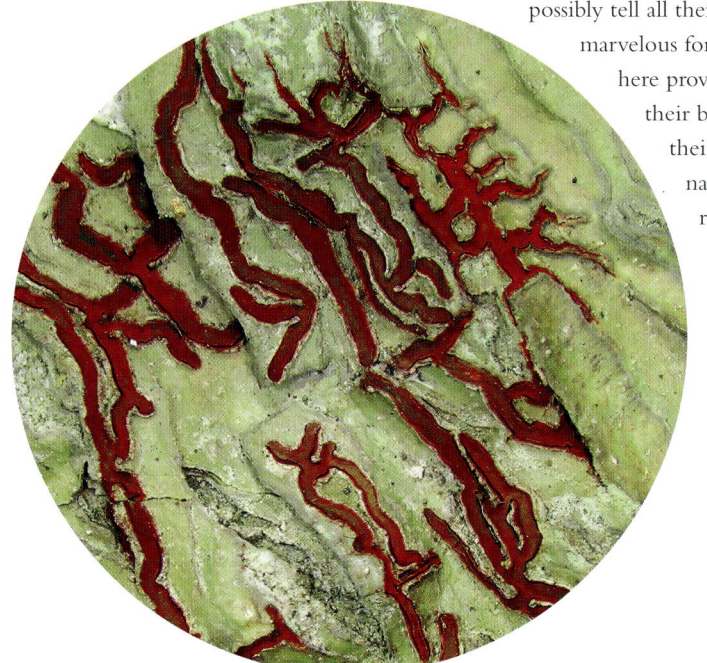

← Script lichens of the family Graphidaceae adorn tree bark and other substrates, mostly in tropical and subtropical regions. Over 2,000 species are known, including the Common Bloodscript Lichen (*Phaeographis haematites*).

→ Orange bush lichens of the genus *Teloschistes* (here *Teloschistes exilis*) hint at the striking color palette lichens have to offer, caused by diverse biochemical compounds often known only from these symbiotic organisms.

THE
ARCHETYPAL
SYMBIOSIS

Early debates about lichens

Lichens have been familiar to humans since antiquity, doubtless under many names. It was the Greek naturalist Theophrastos (371–287 BCE) who introduced the term lichen—in its original Greek, as λειχήν—to refer to patches on olive trees that he assumed affected their growth.

Whether or not these patches described by Theophrastos were lichens as we think of them today, we may never know. What we do know is that in at least two other cases where he did discuss what we today call lichens, he did not refer to them as such. He treated the popular dye lichen *Roccella*, for example, alongside algae, while discussing tree-dwelling *Usnea* lichens among the mosses. His teachings influenced the writings of Pliny the Elder and Pedanius Dioscorides in the first century CE and remained the basis for botanical knowledge of lichens in Europe for almost two millennia.

FOOD AND MEDICINE

The knowledge of lichens in antiquity was carried forward and enriched by the renowned Persian medic and natural historian Ibn Sīnā (Avicenna) (980–1037), who wrote about lichens used in medicine and food in his encyclopedia *al-Qānūn fī al-Ṭibb* (Canon of Medicine). The lichens were referred to as *ushnah*, thought to be the root of the modern name for *Usnea* lichens, and were prescribed for a wide range of ailments. Ibn Sīnā's work was expanded upon two centuries later by the physician Ibn al-Bayṭār (1197–1248) in al-Andalus, in current-day Spain. The exact lichen species these writers describe cannot be interpreted with certainty, but at least some grew on trees and at least one was referred to as *Shaibat al-Ajuz* ("the old woman's gray hair"), echoing modern names for hair lichens.

In many other places across the world, lichens were part and parcel of daily life, particularly in regions rich in them, such as the mountains and the far north. However, there is little preserved evidence of traditional Indigenous knowledge of lichen diversity—beyond a few species used for food and medicine—until the late seventeenth century.

THE STUDY OF LICHEN DIVERSITY

In the late 1600s, several attempts were made to begin a more complete and systematic accounting of species diversity in the broader plant world (lichens were treated as plants during this period). The Englishman John Ray (1627–1705) in his *Historia Plantarum* in 1686 published an account of known lichen species. He may have been the first to introduce the term *Lichen* as a classification category for what we now call lichens, employing a term that had previously been used mostly for liverworts and algae. An even more consequential book, published in 1700, was *Institutiones Rei Herbariae* by Joseph Pitton de Tournefort.

However, neither of these works was as comprehensive or as beautifully illustrated as that of Robert Morison (1620–1683). Morison explored lichens in the Scottish highlands, and in a posthumously published book (1699) he illustrated more than 100 species. Then, in the early part of the eighteenth century, Pier Antoni Micheli and Johann Jakob Dillenius each produced detailed classifications of nearly 300 species, many of them new. Curiously, while the Italian Micheli, like de Tournefort, used the generic

← Page of Ibn Sīnā's *al-Qānūn fī al-Tibb*, the Canon of Medicine, a work that contains one of the earliest references to *ushnah*.

→ *Usnea articulata*, still known as Razi's Moss or Avicenna's Moss in Iran. Ibn Sīnā is thought to have learned about the use of these lichens through his teacher Abū Bakr Muhammad ibn Zakariyyā al-Rāzī, who harked from the Hyrcanian forest region in the north of Persia, and penned the earliest known reference to these lichens as *ushnah*.

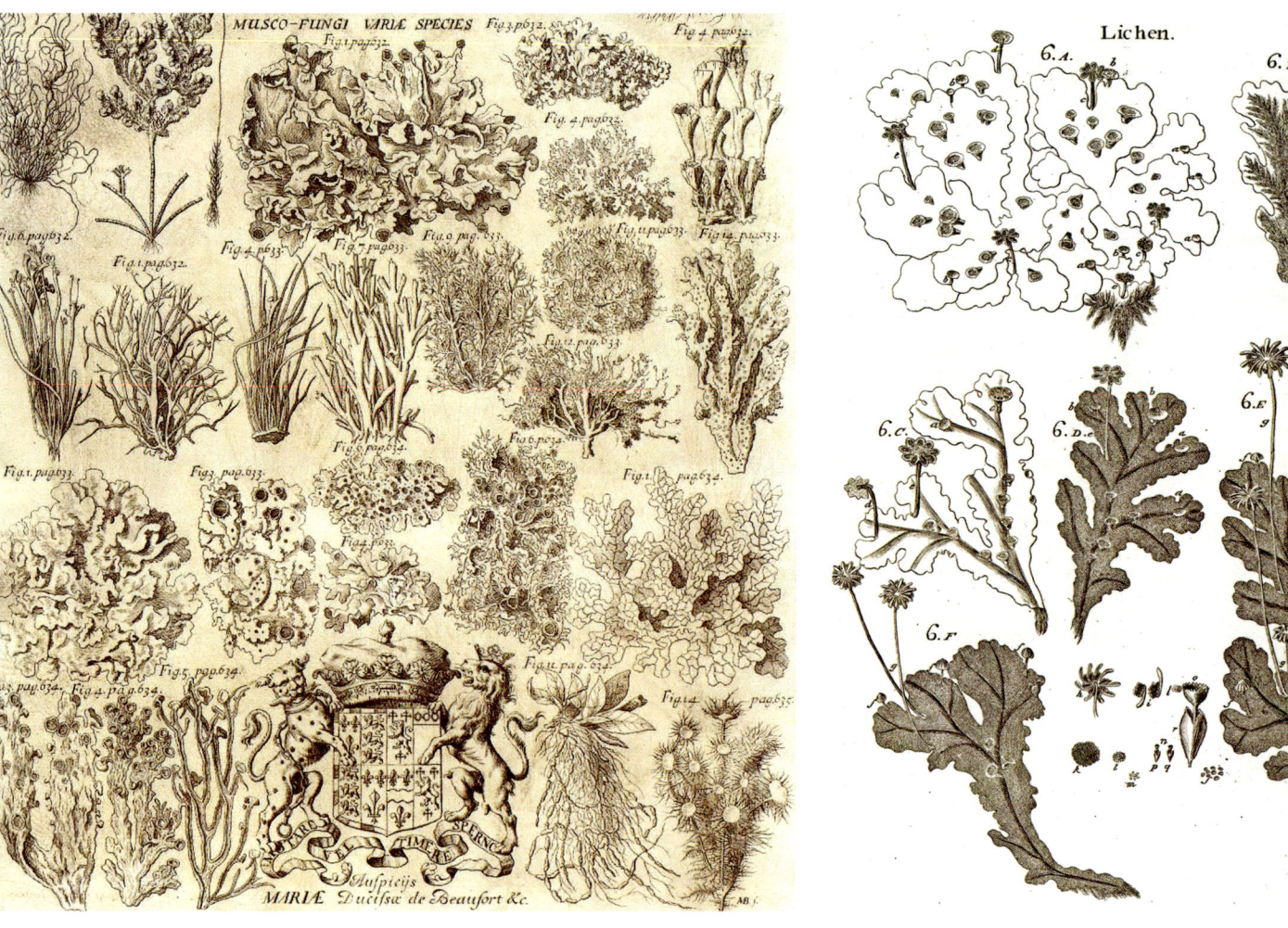

name *Lichen* for all lichens, the German Dillenius treated most of the lichens under the genus names *Coralloides* (mostly *Cladonia*), *Lichenoides* (crustose and foliose lichens), *Usnea* (shrubby and beard lichens), *Byssus* (including leprose lichens), and *Tremella* (including jelly lichens). The genera *Lichen* and *Lichenastrum*, by contrast, he reserved for liverworts. Though it was starting to be used, the term "lichen" was clearly not yet universally agreed nomenclature for lichens in English or Latin, as Morison referred to them as "musco-fungi," or moss fungi, and Dillenius placed them among the mosses.

THE LINNAEAN REVOLUTION

The works of Micheli and Dillenius were influential in guiding the founder of modern botany, Carl Linnaeus, in his classification of lichens. Most researchers before Linnaeus had used more than two Latin words to name species, but in 1753 Linnaeus standardized the system to a tidy two, consisting of genus and species names. His landmark work *Species Plantarum* is now accepted as the starting point for modern scientific nomenclature. Linnaeus is said to have had little interest in lichens, listing only about 80 species in this book, compared to the nearly 300 species recognized by

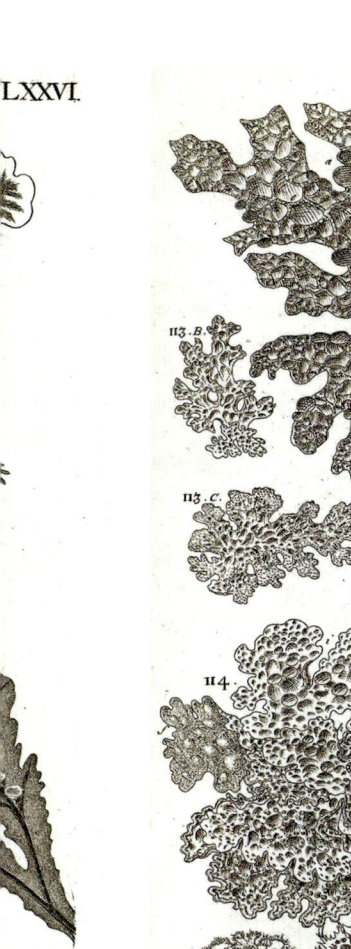

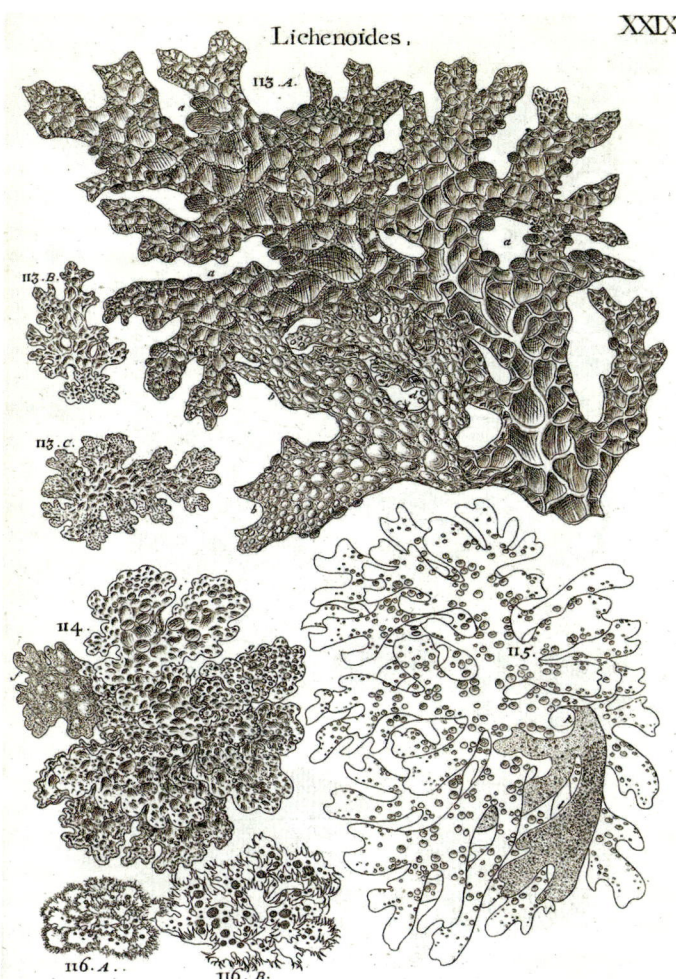

↖ Lichens as "moss-fungi" in Robert Morison's *Plantarum Historiae Universalis* (1699).

↑ Two plates from Dillenius (1741), illustrating his use of the term "Lichen" for thallose liverworts such as *Marchantia polymorpha*, and "Lichenoides" for what we call lichens today, including the Lung Lichen (*Lobaria pulmonaria*).

↗ *Cladonia pyxidata*, the Pebbled Cup Lichen, a representative of the trumpet or pixie cup lichens in the genus *Cladonia*, is one of the species originally recognized by Linnaeus.

Micheli and Dillenius. In contrast to Dillenius, and following de Tournefort and Micheli, he kept the genus *Lichen* for most lichens, but placed them in the algae, alongside seaweeds. Of the 80 names in the genus *Lichen* used by Linnaeus, 67 are still in use today. Commonly used scientific names, such as *Cetraria islandica* for the Icelandic Moss, *Cladonia pyxidata* for the Pebbled Cup Lichen, and *Hypogymnia physodes* for the Tube Lichen, ultimately go back to names in this work, in which Linnaeus listed these species as *Lichen islandicus*, *Lichen pyxidatus*, and *Lichen physodes*, respectively.

Lichens under the microscope

Carl Linnaeus revolutionized scientific nomenclature, setting up a naming system that has largely stood the test of time. This does not mean he was at the cutting edge of contemporary science in every respect. For the smaller organisms he studied, Linneaus took a pass on using the latest technology.

Despite the microscope having been invented in the late 1500s, and entire books having been published on microorganisms prior to 1700 by pioneers such as Antonie van Leeuwenhoek, Linnaeus did not embrace the use of the microscope. This greatly limited his ability to make detailed observations on lichens and other organisms. Spores, for instance, did not feature in his work. It took another half-century before a new generation began to explore the world of lichens through the microscope.

UNLOCKING TINY WORLDS

Linnaeus's last student was a young botanist by the name of Erik Acharius. In 1776, at the age of 19, Acharius completed a thesis on *Hydnora*, a bizarre parasitic vascular plant from Africa that lacks chlorophyll, but he did not continue working with vascular plants, nor on the academic path. Instead, he became the town medical officer and later director of the hospital in the small Swedish town of Vadstena—and here it was that Acharius took up the task of cataloging lichen diversity, armed with more modern scientific tools, and with the ambition to name species on a scale that had not been undertaken before. Acharius used a microscope to document what only a few of his contemporaries had seen, and Linnaeus had not: the tiny world of lichen structures. He described and illustrated fruiting bodies and many other microscopic features in the glossary to his 1803 book, *Methodus Qua Omnes Detectos Lichenes* ("The method by which all lichens are detected"). Seven years later,

in his even more ambitiously titled 700-page tome *Lichenographia Universalis*, he illustrated spores and other fine details that suggest he had access to a microscope of around 400× magnification. Acharius not only lived by the microscopic technique, he also died by it, succumbing to a stroke at the microscope in his garden on August 14, 1819, aged 61, while examining a collection of lichens sent to him from Spain. But Acharius's interest in lichen diversity and use of the microscope lived on, and was taken up by many others.

THE DISCOVERY OF "GONIDIA"

Much of early lichenology was focused on the thrilling task of documenting and giving names to the diversity of lichens. All of the branches of science were undergoing explosive growth at this time, with ground-breaking discoveries in chemistry and physics happening around the late eighteenth and early nineteenth century. For lichens, it did not take long before lines of scientific inquiry moved from cataloging diversity to investigating how lichens work as organisms. One of the pioneers in this field was a German botanist by the name of Karl Friedrich Wilhelm Wallroth (1792–1857). It was with Wallroth's work that science inched one step closer to one of the most profound scientific discoveries involving lichens yet to be made.

What we now know to be algal cells were not mentioned in the earliest works on lichens, at least not as recognizable cells. At most they were afforded mention as a greenish layer or discoloration of the lichen body, the thallus. Some authors even thought this discoloration happened upon exposure of the freshly broken thallus to air. Wallroth changed this in 1825 with a verbose and pompously written two-volume, 722-page opus that both introduced terms and concepts for these cells and ridiculed his predecessors for overlooking them. Wallroth gave the tiny green structures the German name *Brutzellen* ("brood cells") and the Latin term *gonidia* (a term derived from the same etymological root as gonad). He proclaimed, with sentences long even by German standards of the era, that they were "anaphroditic reproduction organs":

← Georg Hoffmann, a contemporary of Acharius, was so enthusiastic about the use of microscopes on lichens that he commissioned a frontispiece depicting cherubic lichenologists conducting field studies for his 1790 book *Descriptio et Adumbratio Plantarum e Classe Cryptogamica Linnæi Quæ Lichenes Dicuntur* ("Description and sketch of plants from Linnaeus's cryptogamic class called Lichens").

This is the first correct explanation of an excellent, meaningful, but, as we have already shown, unfortunately partly overlooked by all authors and partly misjudged part, which we confidently, if also against the claims of all authors, accord the status of an organ, in part; by whose insight, in addition to that of others, with the diagnosis of types of related doubts, the secret of the origin, morphosis and metamorphosis of lichens, can for the most part be deciphered.

→ A plate from Acharius's work *Lichenographia Universalis*, focusing on the reproductive structures of macrolichens in the genera *Sticta* and *Parmelia*, as interpreted at the time. The typically green or blue-green layer easily seen in lichen sections is not depicted, suggesting that Acharius was not able to discern this layer from the cortex.

↓ Reproduction, photosynthesis, or both? Early scientists puzzled over what to make of the green cells inside lichen thalli.

In other words, Wallroth was convinced that gonidia were a key piece of lichen biology, and in this, at least, he would prove to be right. Whether or not they were lichen reproductive organs, however, was about to become the topic of heated scientific debate. Photosynthesis was already a well-known scientific concept by this time, and it did not take long for scientists to note the photosynthetic potential of those green cells. Wallroth's definition of gonidia as asexual reproduction bodies was thus expanded by the Swedish mycologist Elias Fries in 1831 to include another function, carbon assimilation and nutrition. Fries still insisted on the reproductive nature of the gonidia,

however, ascribing to them the regenerative capacity of structures we now call soredia (see page 52).

Throughout this period, lichens were largely treated as a distinct branch of life on the level with plants, seaweeds, and fungi. But the similarity of some lichen structures, especially fruiting bodies, to those of fungi was hard to overlook, and elicited increasingly frequent comparisons. In 1833 the German botanist Friedrich Traugott Kützing claimed to have observed the Yellow Wall Lichen (*Xanthoria parietina*) arising from single-celled algae. Kützing referred to these algae being "woven into" the lichen thallus over time, but he did not make the connection to Wallroth's

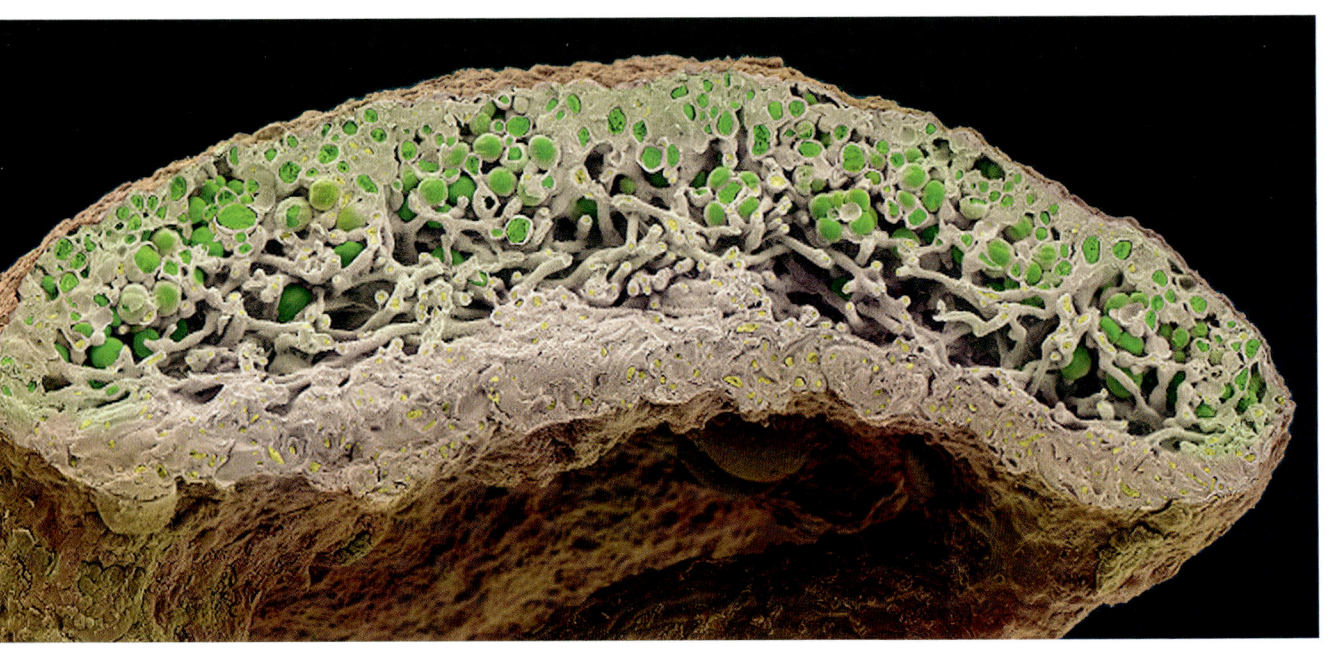

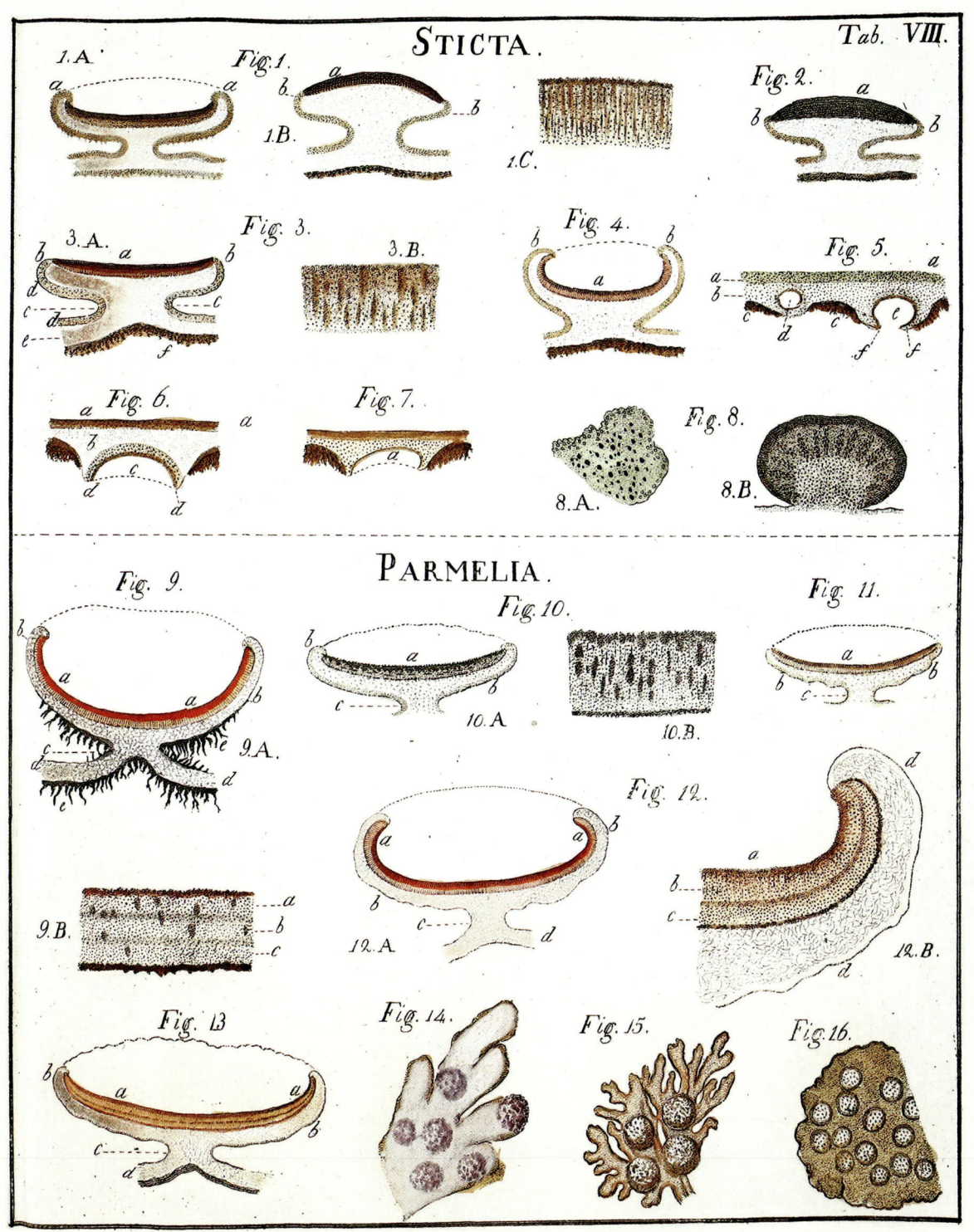

STICTA.

Tab. VIII.

1.A.   Fig.1.   1.B.   1.C.   Fig.2.

Fig. 3.   3.A.   3.B.   Fig. 4.   Fig. 5.

Fig. 6.   Fig.7.   Fig. 8.   8.A.   8.B.

PARMELIA.

Fig. 9.   Fig.10.   10.A.   10.B.   Fig. 11.

9.A.   Fig. 12.

9.B.   12.A.   12.B.

Fig. 13   Fig. 14.   Fig. 15.   Fig.16.

concept of gonidia, and dedicated little text to contemplating what this might mean for lichen biology. The conclusion that lichens were composites of two different organisms—a fungus and an alga—was within grasp, but it would take 36 more years for that bridge to be crossed.

Throughout the 1840s and 1850s, several scientists cultured free-living gonidia from lichens and found that they could even form structures such as free-swimming zoospores that beyond any doubt placed them with other algae. But instead of concluding that they were algae living within lichens, these scientists concluded that the algae that exhibited these behaviors within lichens were in fact not algae at all. To explain their existence, the lichen life cycle was woven into an ever more complicated story that involved reproduction via

complex gonidia that *resembled* algae. As the classification of major forms of life came into sharper focus—with "kingdoms" such as Fungi and Plantae taking center stage—the more problematic it became to have organisms with both fungus-like and plant-like traits. The Occam's razor solution was perhaps as obvious as ever, but it fell to one Swiss botanist, Simon Schwendener, to propose it.

↖↗ One of Schwendener's study objects was the Soil Ruby, involving the fungus *Heppia adglutinata* and cyanobacteria, with a section through a thallus depicted in his 1869 work (opposite).

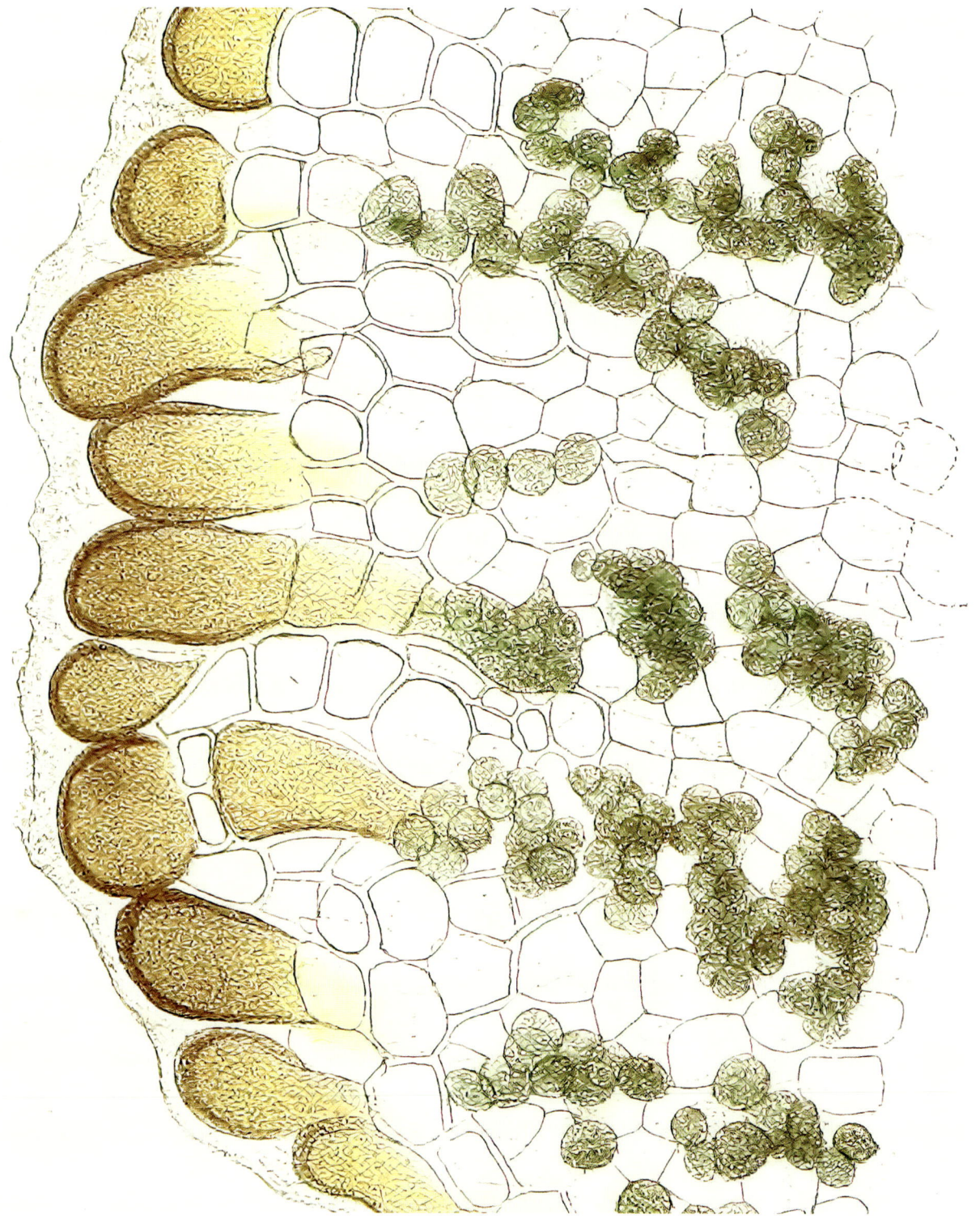

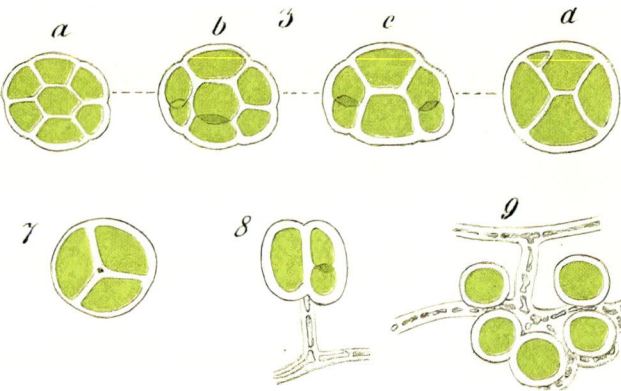

↖↑ Illustrations of algae, and fungi contacting them, from Schwendener's 1860 work *Untersuchungen über den Flechtenthallus* ("Studies on the lichen thallus"), from a species now known as *Alectoria sarmentosa* (left).

→ A false-colored scanning electron micrograph of similar structures in *Ramalina dilacerata* from Alberta, Canada.

↓ Discoverer of the symbiotic nature of lichens: Simon Schwendener.

"GONIDIA" ARE ALGAE

Simon Schwendener (1829–1919) was a 31-year-old professor in Basel when he published a long and beautifully illustrated study on lichen structures in 1860. He produced detailed drawings of gonidia and the cells that enmesh them, and began to think deeply about what they could be. He subsequently replicated the removal and culture of gonidia and, like his predecessors, compared them to free-living algae. Where Schwendener revolutionized science was not in performing an unprecedented experiment but in expressing the previously unthinkable. Schwendener proposed that the lichen was part fungus, part alga. In essence, he advanced the idea that lichens were living chimeras: manifestations of the relationship of one organism feeding the other. Though he did not yet use the term—that would fall to his contemporary Albert Frank—Schwendener introduced the concept of symbiosis, literally, "together life." (For the pedantic, several of the "algae" he studied were in fact cyanobacteria, but they were thought to be algae at the time; page 50.)

Because it overturned a paradigm, Schwendener's work ultimately became a milestone in the history of biology, arguably on a par with Darwin's theory of natural selection. In the following decades, this was not always appreciated, and the resonance of Schwendener's work differed greatly depending on the audience.

WEATHERING THE STORM

What ensued has been called "the war of the lichenologists." The first recipient of the backlash was Schwendener himself. In the first few years after proposing that lichens were chimeras of fungus and alga, he had to endure withering criticism. The worst of it came from leading lichenologists of the time, and it is fair to say this was not an exercise in level-headed hypothesis testing. The Finnish lichenologist William Nylander penned a diatribe entitled *Stultitia Schwenderiana*—the "stupidity of Schwendener."

Meanwhile, in Germany, Gustav Wilhelm Körber ridiculed Schwendener for supposedly cherry-picking a few lichen species for his study, and castigated him for implying that scientists who had studied hundreds of lichens would overlook the relationship of two distinct organisms.

Beyond the insults, it is hard to understand why exactly lichenologists were so incensed by Schwendener's conclusions, other than that they thought that lichens might lose their cherished standalone branch in the tree of life. If Schwendener was upset by this criticism, he did not show it. His explanations for lichen biology were adopted by at least one mainstream textbook within only three years, a development that he remarked, perhaps with a wink at his critics, was "more than I had ever expected."

Recreating lichens in the lab

The ultimate proof of what makes a lichen is to build one in the lab from its component parts. Almost immediately after Schwendener's paper on the dual nature of lichens, scientists set themselves to the task. It turned out not to be so simple.

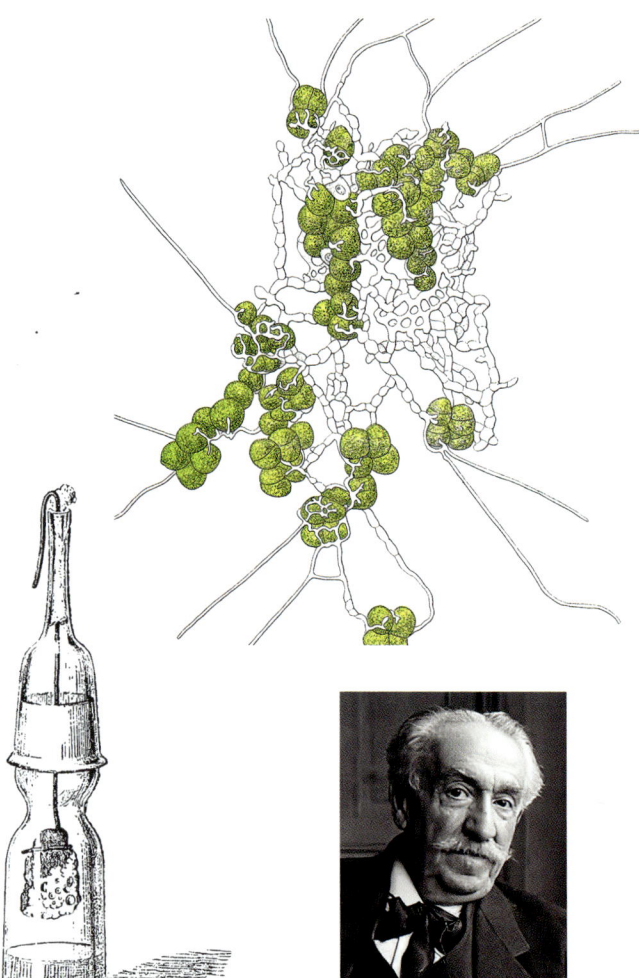

The rush to synthesize a lichen from its component parts—and "prove" its dual nature—coincided with the rising fame of the work of the French microbiologist Louis Pasteur. Among Pasteur's lasting contributions to science was the introduction of a sterile environment for culturing microbes. A decade after Schwendener, the French botanist Gaston Bonnier, familiar with Pasteur's recent success in culturing bacteria, obtained cultures of fungus and alga and in 1889 reported lichen resynthesis. The problem was solved—or was it?

TRIAL AND ERROR

Bonnier was not the first to attempt to resynthesize lichens from their component parts; that distinction goes to the German botanist Maximilian Reess, who attempted a synthesis of a *Collema* lichen in 1871. But Bonnier did achieve some of the best results at the time, using methods that foreshadowed modern microbiology.

← Successful lichen resynthesis? Gaston Bonnier, one of his Pasteur-flacon contraptions, and the results of a resynthesis from the symbionts forming the Yellow Wall Lichen, with the fungus *Xanthoria parietina*.

↗ Lichen resynthesis experiments in petri dishes by Beata Guzow-Krzemińska and Elfie Stocker-Wörgötter, using the Chewing Gum Lichen (*Protoparmeliopsis muralis*) with *Trebouxia* photobionts. The electron microscope reveals the intricate associations of the symbionts in culture.

CLADONIA CRISTATELLA

British Soldier Lichen

Flamboyant ambassadors of the lichen world

SCIENTIFIC NAME	*Cladonia cristatella* Tuck
PHYLUM, FAMILY	Ascomycota, Cladoniaceae
GROWTH FORM	Thallus with squamulose basal part and fertile podetia with brilliant scarlet apothecia
SPECIES IN GENUS	500
HABITAT	On ground and rotten wood
NOTABLE FEATURES	One of the most recognizable and photographed lichens in Canada and the eastern United States

In parts of eastern North America, the British Soldier Lichen, or Scarlet-Crested Cladonia, may be the first lichen many people will learn. People with no interest in lichens whatsoever have been known to stoop down to inspect its scarlet fruiting bodies (apothecia), and it acquired a common English name before most other lichens in this region. Can it tell a story?

The genus *Cladonia* is one of the largest genera of lichen fungi worldwide, and it forms a plethora of thallus forms in association with green algae from the genus *Asterochloris*. *Cladonia cristatella* is one of numerous species worldwide that produce scarlet apothecia at the tips of their vertical thallus parts (podetia), and it is one of the most spectacular.

The function of the reddish pigment is not well understood, but in at least one species it has been postulated to be induced by lichenivorous mites, perhaps as a defense compound. However, cladonias with brown apothecia appear to fare just fine, and if mites are responsible for the pigmented compound in *C. cristatella* apothecia, they are certainly few and far between. Not all science is settled.

Cladonia cristatella was one of first lichens to be given a common name in American natural history books. Nina Marshall, in a popular guide to lichens published in 1907, called it Scarlet-Crested Cladonia, a nod to the Latin name (*crista*, crest). The name British Soldiers appears to date from 1932, in a report on lichens along a trail from Tomkins Cove to Bear Mountain, New York:

These "British soldiers," so-called because their scarlet apothecia suggested the bright uniforms of the redcoats in the American Revolution, were to be seen along the route followed in October 1777 by the British force which climbed over Dunderberg Mountain to storm Forts Clinton and Montgomery.

→ Rolling out the red carpet: *Cladonia cristatella* often caps old stumps in a splash of color.

NEPHROMA ARCTICUM

Arctic Paw Lichen

Triumph of the tripartite

SCIENTIFIC NAME	*Nephroma arcticum* (L.) Torss.
PHYLUM, FAMILY	Ascomycota, Nephromataceae
GROWTH FORM	Large foliose lichen with broad, opaque greenish lobes, with a dominant green alga and internal colonies of cyanobacteria (internal cephalodia)
SPECIES IN GENUS	36
HABITAT	On sheltered ground in boreal to Arctic regions
NOTABLE FEATURES	Consistently harbors two photobionts, one of which fixes nitrogen

The large majority of lichens are a symbiosis with a characteristic and fairly predictable ratio of microbial players. There is usually one fungus, which makes up the bulk of the lichen, and one alga or cyanobacterium, often at a ratio of about one cell for every ten fungal cells. Not so the "tripartite lichens."

In most lichens, the two main players are accompanied by various other microbes—other fungi and bacteria—in smaller ratios. But tripartite lichens contain two photobionts—usually a green alga in large parts of the thallus and cyanobacteria in special receptacles called cephalodia, or cyanobacteria in separate lobes or thalli, or even both, as in the Arctic Paw Lichen (also called the Arctic Kidney Lichen).

Why tripartite? Studies suggest that such lichens are a relatively recent innovation, and that the green alga keeps the lichen supplied with sugar alcohols, important for preserving the lichen through its daily drying routine, while the cyanobacterium provides glucose, a valuable energy source.

It was long thought that cyanobacteria in tripartite lichens only fix nitrogen and not carbon, but recent evidence suggests they do both. This dual function might just be the key as to why many tripartite lichens achieve above-average sizes, since nitrogen can be limiting in many places lichens grow. Put simply, fungi simply run out of nitrogen to build cells with, but if a source is built in, the constraint is gone.

The Arctic Paw is one of the lichens that achieves considerable sizes in places not exactly awash with nutrients. In many cases, it is even larger than surrounding vascular plants and smothers the mossy boulders on which it grows.

→ *Nephroma arcticum* has green algae as its main photobiont and additionally cyanobacteria either in small, darkened cephalodia visible on the thallus surface, or rarely, as in this photo, in separate lobes or thalli distinguishable by their dark color.

MASTODIA TESSELLATA

Seaweed Lichen

Seashore lichen in which the symbiont tables have been turned

SCIENTIFIC NAME	*Mastodia tessellata* (Hook. f. & Harv.) Hook. f. & Harv
PHYLUM, FAMILY	Ascomycota, Verrucariaceae
GROWTH FORM	Contorted, dark green foliose thalli dotted with black perithecia
SPECIES IN GENUS	1
HABITAT	On seashore rocks at the high tide line
NOTABLE FEATURES	An "inverted lichen" in which the structure is provided by the alga

It is hard to know what the first lichenologists thought upon encountering the lichen we now call *Mastodia*, but they might be forgiven for seeing a resemblance to jelly lichens (such as *Enchylium tenax*; pages 74–75).

Detailed study of *Mastodia* has shown it to be quite peculiar. Instead of the regular saucer-shaped fruiting bodies called apothecia, it has embedded flask-shaped structures called perithecia, which look like tiny black pimples. Perithecia are also found in some groups of crust lichens, so this is not in itself unusual, but in *Mastodia* the dominant symbiont—the one that forms the bulk of its thallus—is not a fungus but an alga. A seaweed, to be more precise.

That seaweed goes by the name of *Prasiola borealis*. It is not a large seaweed—most are under ¾ in (2 cm) across when fully developed—and its freshwater relatives are harvested for food in Japan (as *kawa-nori*). Sometimes it grows side by side with darker-bladed *Prasiola* and these, by dint of having a fungus, are classified as lichens. *Prasiola* and its *Mastodia* lichen form are found atop bedrock and boulders on the far northern and far southern eastern shores of the Pacific Ocean, around the high tide lines, as well as in cold coastal habitats of the Southern Hemisphere and Antarctica.

Aside from turning the seaweed black and making it look especially crumpled, the fungus does not appear to harm the *Prasiola* alga. But it is unclear what the seaweed gets from the deal.

→ *Mastodia tessellata* infects the seaweed *Prasiola borealis* on a high-tide rock outside Sitka, Alaska.

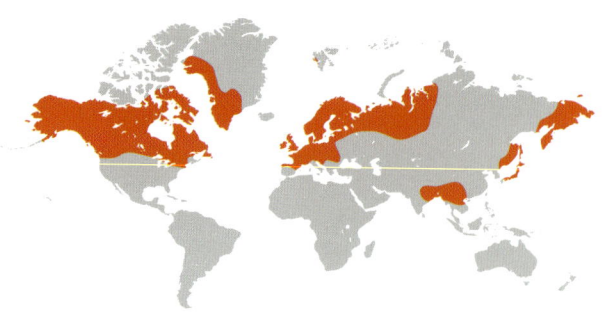

Arctic Mushroom Scales

A lichen and a mushroom in one

SCIENTIFIC NAME	*Lichenomphalia hudsoniana* (H. S. Jenn.) Redhead *et al*
PHYLUM, FAMILY	Basidiomycota, Hygrophoraceae
GROWTH FORM	Thallus of round scales close to substrate; fruiting body a mushroom
SPECIES IN GENUS	16
HABITAT	On decomposing organic matter in mountain and Arctic tundra
NOTABLE FEATURES	A lichen in which the fungus reproduces via mushrooms

Most lichens involve a fungus in the phylum Ascomycota, and the thallus looks like a leaf, shrub, or scale—with the fruiting body, if present, forming a disk-like apothecium. However, some fungi in the phylum Basidiomycota can also get in on the lichen action. And in a few of these, the outcome can be surprising.

Lichenomphalia is a genus of fungi in the family Hygrophoraceae. Members of this group include colorful mushrooms called waxcaps in the genus *Hygrocybe*, which inhabit forest soils and live in obligate association with tree roots. The fungal family is, however, a hodgepodge of root-associated and lichen lifestyles, among which *Lichenomphalia* is just one of eight genera (another is *Cora*; page 188).

In *Lichenomphalia*, the lichen thallus—in which the fungus consorts with an alga from the genus *Coccomyxa*—takes the form of ear-shaped scales that one could be forgiven for mistaking for those of a *Cladonia*. The fruiting bodies are, however, unmistakable: they are bona fide mushrooms.

Lichenomphalia has been the subject of "resynthesis" studies (page 22), in which symbionts are brought together to form a new thallus *in vitro*. Such experiments are notoriously difficult, and in the case of this lichen the fungus was especially recalcitrant. The Austrian lichenologist Elfie Stocker-Wörgötter pulled it off by designing special growth media for the fungus. But a quick result it was not: it took over a year and a half for a single thallus lobe to develop.

→   The Arctic Mushroom Scale in its natural habitat; in the absence of the characteristic mushrooms, it can be recognized by its rounded scales.

Yellow Wall Lichen

Small but streetwise

SCIENTIFIC NAME	*Xanthoria parietina* (L.) Th. Fr.
PHYLUM, FAMILY	Ascomycota, Teloschistaceae
GROWTH FORM	A bright yellow foliose leaf lichen with orange apothecia
SPECIES IN GENUS	10
HABITAT	Ubiquitous in natural and urban areas, on trees, rocks, and walls
NOTABLE FEATURES	Bright yellow-orange in the sun, more grayish in the shade

The Yellow Wall Lichen is one of the most commonly encountered lichens in the towns and countryside of Europe. Its Latin name, *parietina* (from *paries*, wall), is a nod to its occurrence on walls, but in truth *Xanthoria parietina* occurs on almost any available stable substrate.

Xanthoria parietina is a synanthrope—an organism that lives in close association with people, a characteristic that distinguishes it from most other lichens, which have their own reasons to be misanthropes. How did one species become so closely associated with human habitation? Another one of its common names, Maritime Sunburst Lichen, gives a clue. In unspoiled nature, *Xanthoria parietina* can be found on trees and rocks along seashores, where high levels of nitrogen and phosphorus are delivered in the form of seabird droppings. In ecological terms, this is not that unlike environments enriched by farmyard manure. From the lichen's perspective, the increasing size and frequency of agricultural feedlots and the emission of nitrous oxides by automobiles has ensured that the human *Umwelt* itself has turned into one giant seabird roost.

The Yellow Wall Lichen is also one of the few lichens to have become demonstrably invasive. Native to Eurasia and the north coast of Africa, it has been known to occur in eastern North America since the late 1820s, and subsequently spread into California. Its march is ongoing: Yellow Wall Lichen turned up on North America's northwest coast in the 1980s, and in recent years citizen scientists have been tracking its spread from the Seattle–Vancouver area ever farther inland, mostly hitching a ride on nursery trees in urban landscaping. In these areas it has yet to jump onto native trees or even concrete, as it does in Europe, but given the adaptability of this lichen, this might only be a matter of time.

Terminal cells

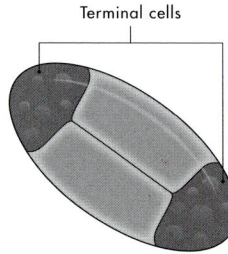

Inside the apothecia

The spores of *Xanthoria parietina* have a peculiar shape, with two cells separated by a thick wall but connected by a thin channel, also typical of other members of the family Teloschistaceae.

→ A familiar site in western Europe and increasingly also in other parts of the world, the Yellow Wall Lichen abounds in nutrient-rich synanthropic habitats, such as the walls of buildings, and trees in parks and along roads.

THE
PLAYERS

Lichen fungi and their relatives

Life on Earth was single-celled for a long time. Several important milestones mark the path to where we are now, and symbiosis—the "living together" and toolbox-sharing of two organisms—features prominently in this lineup. Scientists are now convinced that the switch to cells with nuclei—eukaryotes—happened around 2.5 billion years ago, when an early unicellular organism in a group called Archaea enveloped, but did not digest, a bacterial cell.

This new cell—nicknamed by scientists LECA, for Last Eukaryotic Common Ancestor—is thought to have acquired a powerful advantage over its predecessors. Not only could it carry on with its archaean predator machinery, it could tap into the bacterial baggage it had incorporated to generate adenosine triphosphate, or ATP, a major source of energy for the cell. The twinning of these two toolboxes propelled LECA into places where no archaean had gone before. Passed from generation to generation, the captured bacterium would eventually evolve into what we call mitochondria.

Something similar happened again around a billion years later. Some of the eukaryote descendants of LECA took up another group of nifty bacteria that had been on the scene for a while: cyanobacteria. Cyanobacteria had previously evolved the ability to fix carbon through photosynthesis, giving them the luxury of not having to eat other organisms to make a living. The green descendants of this event are now called plants.

THE RISE OF PROTOLICHENS?

So how did we get from there to lichens? Fungi (and animals) are descended from one of the eukaryotic lines that did not go green. Some of the earliest fungi have been estimated to have arisen around one billion years ago, at the end of the Mesoproterozoic era, probably in shallow marine environments when much more of the Earth's surface was covered in water than it is today. The Mesoproterozoic saw some important advances in the evolution of life, including sexuality and multicellularity. At the same time, fungi had to contend with challenges such as low oxygen levels.

→ *Ourasphaira giraldae*, an enigmatic Late Mesoproterozoic to Early Neoproterozoic fossil considered the oldest fungus-like fossil known to date, from around one billion years ago.

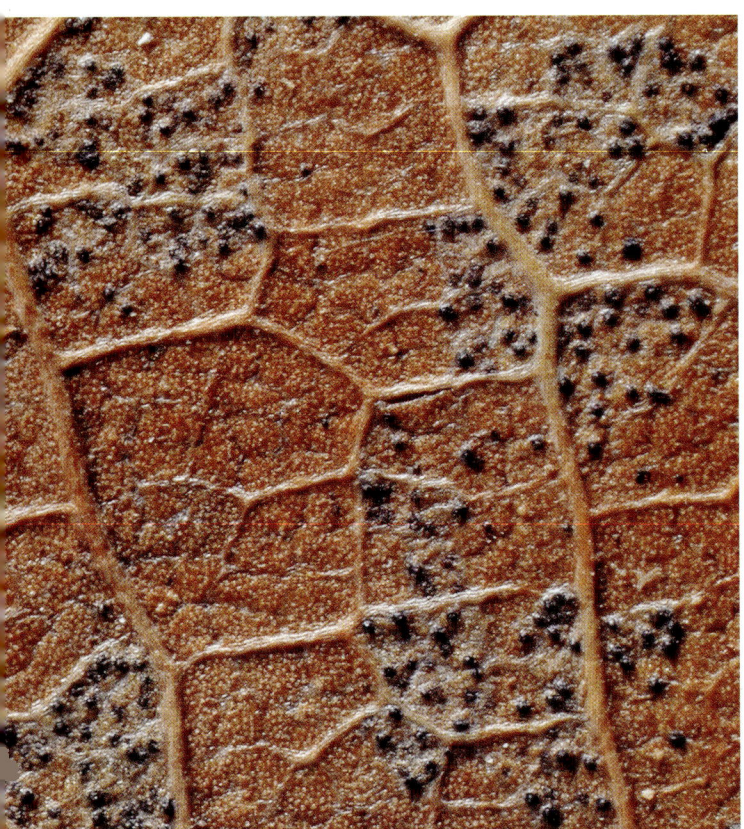

Keeping tabs on our timeline, Mesoproterozoic fungi would have been in the aqueous company of both a well-established veteran community of cyanobacteria that had already been around for a respectable billion years, as well as some upstart plant prototypes in the form of early green algae that were themselves probably not more than a few cells in size. What kind of lifestyle these early fungi led, and how the terrestrial fungi we know today got started, is the matter of much debate. One intriguing hypothesis is that they began appearing on ancient seashores as lichens—the so-called "protolichen hypothesis." Under this scenario, the majority of modern fungi we are familiar with, from yeasts to mushrooms, would be the descendants of fungi that got their first leg up, so to speak, in a lichen-themed amphibious landing vessel.

As intriguing as this idea is, we do not have a lot of data to back it up. Fossils of fungi and lichens do not preserve well and are exceedingly rare, especially this far back in time. Even with the most sophisticated DNA-based reconstruction techniques, it is difficult to be certain about evolutionary events that far back

↖ *Mycosphaerella punctiformis* is a saprotrophic microfungus occurring on dead or dying leaves, especially of oaks, but is also found in healthy leaves. It is closely related to *Cystocoleus ebeneus*, a fungus involved in tiny, filamentous lichens. The genus *Trichoglossum* (here *T. hirsutum*) likely forms mycorrhizae with plants, but is part of a clade containing the likes of the Candleflame Lichen (*Candelaria concolor*), certain pin lichens, and certain cyanolichens.

↗ The mushrooms *Hydnum repandum* and *Arrhenia spathulata*, the first mycorrhiza-forming, the second associated with mosses, are close to the lichen-forming genera *Multiclavula* and *Dictyonema*.

Depending on your perspective, the word "alga" may conjure images of nori, kelp, or toxic blooms, but the algae that occur in lichens are none of the above. They are single-celled, or simple multicellular thread-like organisms. When they occur as free-living organisms, we perceive them as powdery green coatings or minutely hairy orange fuzz on tree bark and rock, if we notice them at all. Like fungi, these algae interact with their surroundings by secreting and taking up biomolecules through their cell walls. Where fungi and animals are heterotrophs, algae are almost always autotrophs—deriving their food from sunlight, water, and carbon dioxide.

SECRET CRAVINGS

But are they *only* autotrophs? It has long been known that some algae, including those found in lichens, grow as well or better in total darkness as in the light, and for prolonged periods of time. This is something that cannot, for the most part, be said of plants. It suggests that, while these algae can live autotrophically, making their own food from water and CO_2, they have a secret back-up plan. Indeed, it is now understood that algae, in addition to performing photosynthesis, can use other sources of carbohydrates for food much as fungi do, and when they do, they are at least partially heterotrophs, just like fungi.

Organisms that switch in such a way between different nutrient modes are called mixotrophs. One curious example is the algal genus *Cephaleuros*, which behaves more like a fungus than an alga, living parasitically on plants, including important fruit trees such as citrus. *Cephaleuros* is also the photobiont of the lichenized fungus *Strigula*.

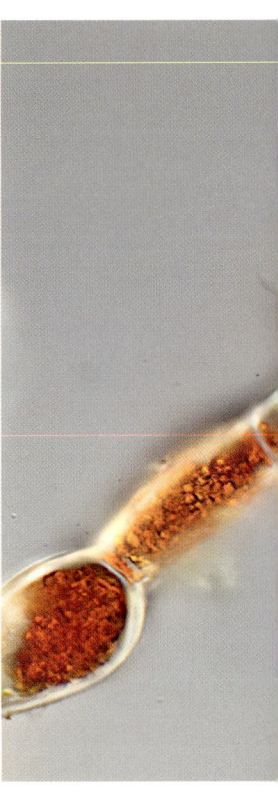

ALGAL REPRODUCTION

Algal sex, like fungal sex, is a cryptic business, and it is not often they are caught *in flagrante*. In fact, algae are thought to rarely reproduce sexually when in lichens, so much so that until 30 years ago lichen textbooks stated matter-of-factly that they were incapable of sex. (Notable exceptions are the aforementioned *Cephaleuros* and the genus *Phycopeltis*, which often produce sporangia in healthy lichens found on leaves in the tropics.) Since then, many lichen algae have been cultured *in vitro* in labs.

In contrast to the fungus, algae almost never reproduce sexually when inside a lichen, and to say these single-celled organisms are difficult to observe in nature when outside of a lichen is an understatement. Nonetheless, they can sometimes be observed reproducing sexually in a petri dish. When a lichen algal cell is feeling frisky, it begins to internally differentiate into specialized structures called zoospores that bear flagella at one end. These are basically little motile whips that can be wiggled to move the cell around, most effectively in water. Algae, like fungi,

↖↗ *Trentepohlia* algae (left and middle left, with close-up middle right) are the only eukaryotic lichen photobionts readily visible when they occur outside lichens. Lichens containing these algae—such as *Thelotrema lepadinum*, the Bark Barnacle Lichen (right)—often have specific habitat requirements.

come in different mating types, and with luck one of these flagellum-whipping zoospores might meet another of the opposite mating type. They fuse, lose their flagella, and grow into a new algal cell.

LICHEN ALGAE OUT AND ABOUT

Lichen algae grow well without having fungi around, and most are thought to also live outside of lichens in nature. But where do they live when they are not in lichens? The Austrian phycologist Elisabeth Tschermak-Woess set out to find free-living lichen algae over 45 years ago, and found them on rock and on the bark of various tree species. Lichen algae are generally "aeroterrestrial" algae, meaning that they occur outside water, and exposed to the air, in terrestrial ecosystems. But even so, finding free-living algae (aside from the flashy orange ones—page 62) is a tricky business, and not something that your local iNaturalist community

can help with on Saturday-morning outings with hand lenses. Detection of single-celled algae requires careful swabbing and culturing, and then constant babysitting to pick out algal cell colonies and prevent them from being overgrown by microbial contamination.

In recent years, researchers have resorted to sequencing characteristic DNA signatures from the environment to get a handle on where these elusive microbes occur when they are not in lichens. We still do not have a definitive answer, but some of the most intriguing results show that lichen algae most definitely are floating about in the air we breathe. In fact, both Japanese and French researchers have shown that lichen algae are among the most common microbes in snow and rain—suggesting that precipitation droplets may well even nucleate around these microbial specks.

Cyanobacteria: metabolic powerhouses

For most of the history of modern biology, cyanobacteria were not universally recognized to be bacteria. Instead, they were referred to as "blue-green algae," and in some languages they continue to carry this name.

The slow migration of cyanobacteria in the classification system from the plant kingdom to bacteria started with an 1853 proposal that they were more closely related to non-blue-green bacteria than to algae. This idea was only fully accepted with the application of biochemical techniques and a formalized definition of bacteria. It was not until 1962 that the Canadian microbiologist Roger Stanier proposed a new name, "cyanobacteria," and this was not fully accepted in bacteriology textbooks until 1974. Old habits die hard.

Early scientists can be forgiven for classifying cyanobacteria as algae, and indeed just like the mitochondria discussed at the beginning of this chapter, the chloroplasts that are the photosynthetic hubs of algal cells are now generally accepted to descend from an ancient cyanobacterial symbiont. That being said, algae and cyanobacteria could hardly come from more distant branches of the tree of life.

Cyanobacteria are true bacteria, and their cellular organization is completely unlike that of algae. They also do not reproduce sexually, instead reproducing by specialized vegetative outgrowths called hormogonia. In contrast to green algal symbionts, most cyanobacteria found in lichens are filamentous, even if in the lichen thallus this organization may be modified to resemble single-celled or colonial organisms.

FIXING NITROGEN

Cyanobacteria serve many functions in ecosystems on land and in the sea, but most can do one thing outstandingly well: fix atmospheric nitrogen. All of life needs nitrogen as a building block of basic molecules like amino acids, and in many places on Earth it is scarce. Cyanobacteria possess a special enzymatic pathway to convert nitrogen gas out of the air into molecules such as ammonium that are more chemically reactive, and thus more accessible to other forms of life (page 200). The importance of this process is so central to life on Earth that cyanobacteria are credited with playing a role in the early colonization of land around 2.6 billion years ago.

The nitrogen-fixing abilities of cyanobacteria make them attractive partners in symbiosis. Many organisms, from cicadas to plants, have evolved close partnerships with cyanobacteria. One of the groups of cyanobacteria most commonly involved in lichens is the genus *Nostoc*. *Nostoc* species usually form chains when living in lichens, and are typically embedded in a gel of their own making.

A common misunderstanding about *Nostoc* in lichens is that they are always involved in nitrogen fixation. The cells of cyanobacterial filaments essentially divide labor, with special enlarged cells called heterocysts engaged in nitrogen fixation but not photosynthesis, and smaller cells engaged in photosynthesis but not nitrogen fixation. The proportion of heterocysts in a *Nostoc* photobiont can be used as a proxy to estimate how much nitrogen fixation the cyanobacterium does in the lichen, if indeed it is fixing nitrogen at all.

→ *Nostoc* (A), *Rhizonema* (B), and *Stigonema* (C) are the most common cyanobacterial lichen photobionts. *Stigonema* is found as the primary photobiont in *Spilonema* lichens (here *S. paradoxum*, D).

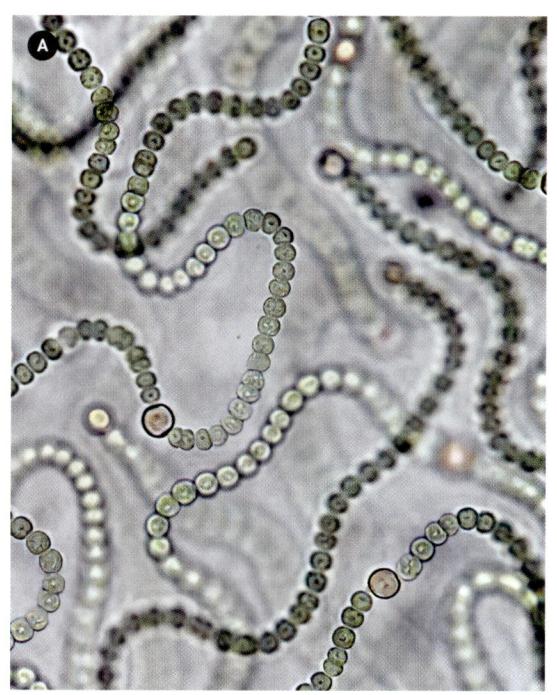

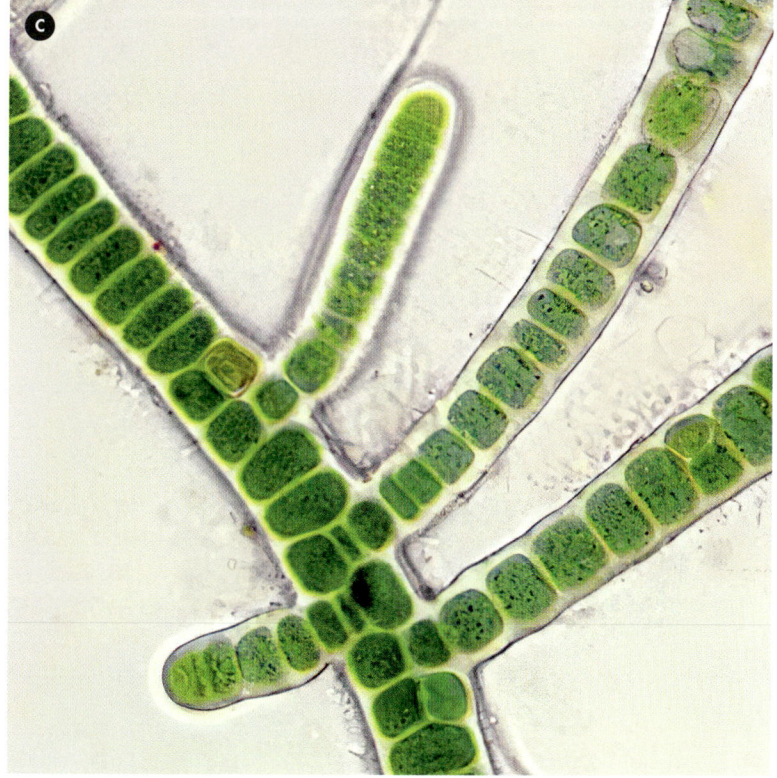

A lichen life:
from cradle to grave

Understanding the biology of individual symbionts is key to piecing together the puzzle of lichen symbiosis. Their individual needs must be met. But when they join in making a lichen, many things become possible that neither could accomplish alone.

↓ Lichen fungi reproducing like fungi. The crust-lichen-forming *Astrothelium megaspermum* disperses its sexually produced spores, which then need to find a new *Trentepohlia* alga to establish another lichen.

So far we have discussed the life cycles of the symbionts, but we haven't especially focused on the lichen itself. How does the lichen, which does not resemble any of its component parts, come to be? One useful metaphor, proposed by the Canadian microbiologist Ford Doolittle, is that of a song sung by a choir with a variable cast of voices. The song remains largely the same—though not exactly—even if individual singers are swapped out, and the song itself can evolve. The rotating door of fungal and algal (and other) symbiont composition makes new lichen songs possible, which can be advantageous to exploring new ecological possibilities. But lichen symbioses have also evolved a way to temporarily lock down the existing singer combination and take some of the luck out of the equation.

MAINTAINING THE LICHEN SYMBIOSIS

Many lichens forgo fungal sexual reproduction and produce just-add-water go-packs that have everything needed to make a new lichen at a fraction of the cost of sexual reproduction. One of the most common versions of go-pack is called the soredium (plural: soredia), in which a few algal or cyanobacterial cells are enmeshed by a small number of fungal hyphae (see also page 136). Soredia can also carry other symbionts, such as yeasts or bacteria. The success rate of lichen formation from soredia is very high in lab experiments, which suggests that it is in nature as well.

A variation on the soredium theme is the isidium (plural: isidia), in which the package is enclosed in a secreted layer of polysaccharides and thus appears to have a hard shell (see page 136).

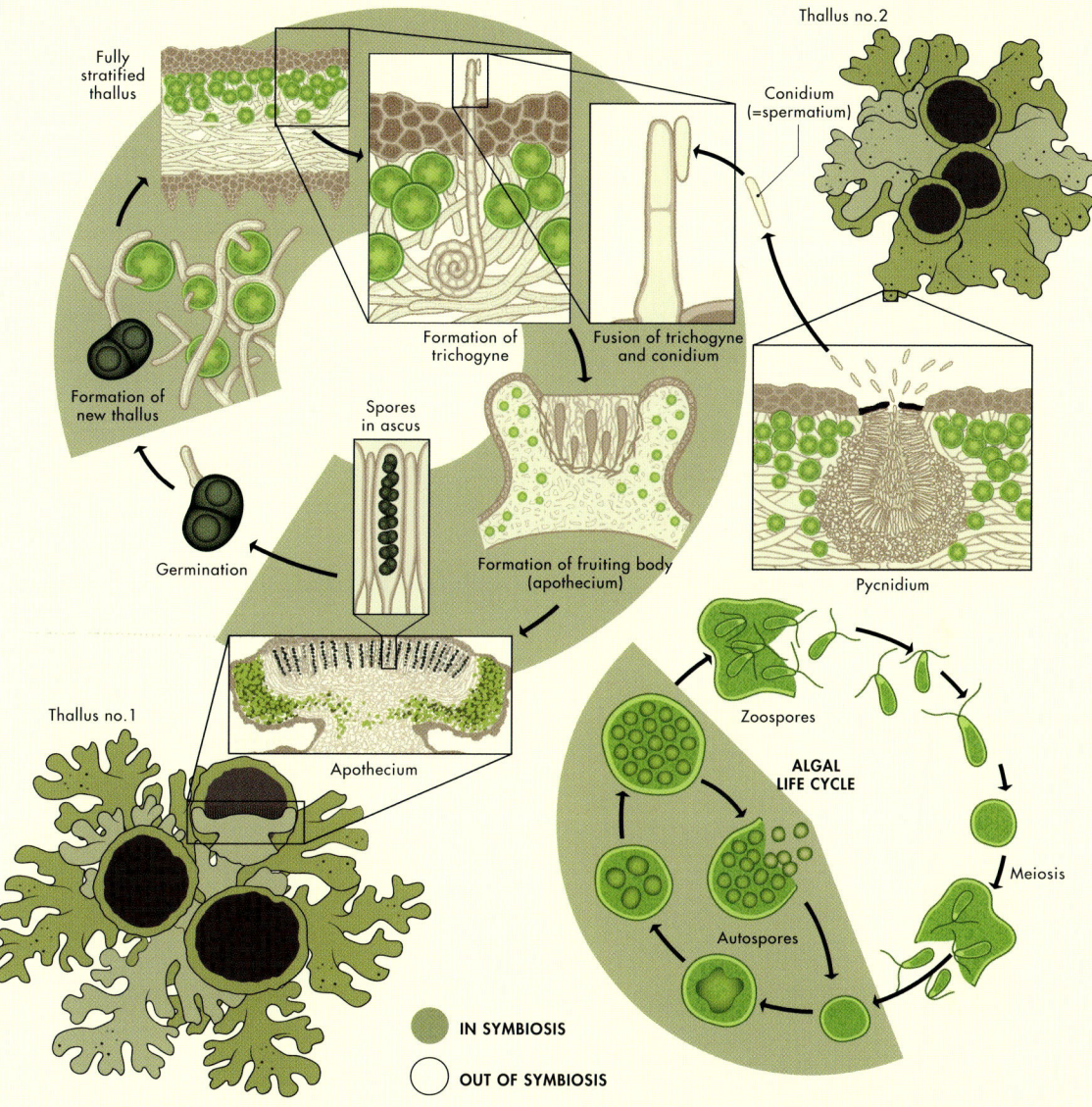

Fully stratified thallus

Formation of trichogyne

Fusion of trichogyne and conidium

Thallus no.2

Conidium (=spermatium)

Pycnidium

Formation of new thallus

Spores in ascus

Germination

Formation of fruiting body (apothecium)

Zoospores

ALGAL LIFE CYCLE

Meiosis

Autospores

Thallus no.1

Apothecium

● IN SYMBIOSIS

○ OUT OF SYMBIOSIS

Isidia likely carry even more microbial cargo and probably retain water better than soredia, but have the apparent disadvantage that they are quite heavy as propagules go. Numerous lichens threatened with severe decline or extinction reproduce with isidia. This might be because they cannot readily disperse to new habitats—even if nearby habitat is suitable, it cannot be reached.

Soredia and isidia make it possible not only for fungus and photobiont to co-disperse, but also lichens carry other symbionts such as bacteria and secondary fungi that hitch a ride on thallus propagules; these may play important, if still poorly understood, roles in lichen establishment.

Sexual life cycles of common lichen symbionts

The fungus (scheme top left) of a sexual *Physcia stellaris* lichen starts life as an ascospore that germinates outside of symbiosis and, once it has found a compatible alga, forms a new thallus. Sexual reproduction initiates when a microscopic trichogyne from this fungus is fertilized by a conidium from a thallus of a compatible mating type. A *Trebouxia* alga (bottom right) meanwhile reproduces sexually outside of symbiosis, but can also reproduce asexually while in symbiosis, using autospores.

Lichen bacteriobionts

In the early 1920s, an Italian microbiologist by the name of Maria Cengia Sambo was culturing cyanobacteria from jelly lichens when she encountered growths of much smaller bacteria on her culture plates. Another key player in lichens entered the picture, but it would be decades before the discovery would begin to be widely appreciated.

Cengia Sambo concluded that the bacteria she cultured could fix nitrogen (which indeed many bacteria other than cyanobacteria can do). She was clearly excited by the discovery, and convinced they were part and parcel of the lichen: "*In simbiosi con l'alga stessa e col fungo,*" she wrote, "in symbiosis with the alga itself and with the fungus." She was not the only one excited by the discovery. Soviet researchers, in particular, followed up with numerous studies from the 1930s through to the 1980s, trying to nail down more evidence of the role of these "other" bacteria in lichens.

UBIQUITOUS BACTERIA

We now know that almost all lichens carry these bacteria. As with us humans, bacterial cells can sometimes outnumber all the other cells put together. In lichens, the most common bacteria worldwide come from just four small groups, treated in the bacterial classification system as families. The most common single line of bacteria of these is called *Lichenihabitans*, because, well, it inhabits lichens.

Despite Cengia Sambo's start a century ago, it's still early days in the science of lichen bacteria. It appears likely that many of the pioneering studies did not, in fact, tap into the bacteria we now know to be ubiquitous in lichens, and these, in turn, have rarely been extracted in culture. We only know they exist in so many lichens because when we grind up lichens and extract their DNA, the DNA of the four bacterial families is very often present. While we still don't have experimental evidence on what they do, their DNA has been helpful in this department too, giving many hints.

← A pioneer in lichen microbiology, Maria Cengia Sambo (1888–1939) isolated numerous bacteria from lichens and hypothesized about their function in the symbiosis.

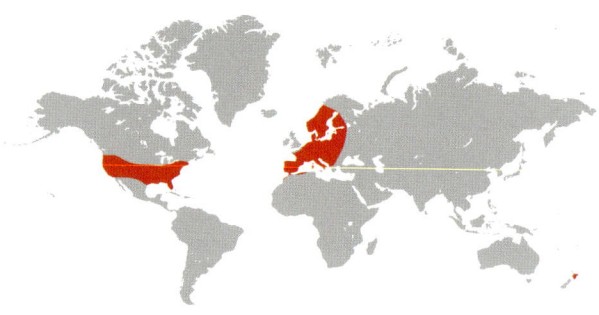

Lynn Margulis's Trebouxia

A photobiont of large and common lichens

SCIENTIFIC NAME	*Trebouxia lynnae* Barreno
PHYLUM, FAMILY	Trebouxiophyceae, Trebouxiaceae
GROWTH FORM	Single cells that reproduce quickly to form a green mass, but not usually visible in nature
SPECIES IN GENUS	27 named, but many are not yet named
HABITAT	Extremely common in terrestrial environments
NOTABLE FEATURES	One of the most common lichen photobionts

Best not try to carry out a survey of lichen photobionts armed with a hand lens and cell-phone camera. Most are single-celled, and identifiable only on the basis of internal cell structures and DNA sequences. As a result, the classification of photobionts has lagged behind that of the fungi involved in lichens, and even some of the commonest species are still being given scientific names.

Break open the thallus of a lichen and squint really hard, and you might make out an extremely thin green line sandwiched between the upper and lower fungal layers. It's the world's thinnest burrito, a fungal tortilla with algal filling. Better yet, use a standard light microscope or a scanning electron microscope (page 21), and you will get a sharper view of round cells wrapped in fungal nets or suspended at the ends

of what look like fungal suction cups. More often than not, these cells are members of the genus *Trebouxia*, one of the most common groups of photobionts worldwide.

A decision to study *Trebouxia* is a conscious decision to become a microbiologist. The methods are much the same as those one might use to culture bacteria from a mouth swab in a hospital, and analyzing the resulting growth requires patience and a well-stocked lab. The classification of *Trebouxia* is a technical business, and many species have yet to be given formal scientific names. One of the more common species in large foliose and fruticose lichens was named as recently as 2022 by a team led by the Spanish researcher Eva Barreno. She named the species for Lynn Margulis, one of the leading figures of symbiosis research and the person who discovered that mitochondria are descendants of intracellular bacterial symbionts.

→ *Trebouxia lynnae.*

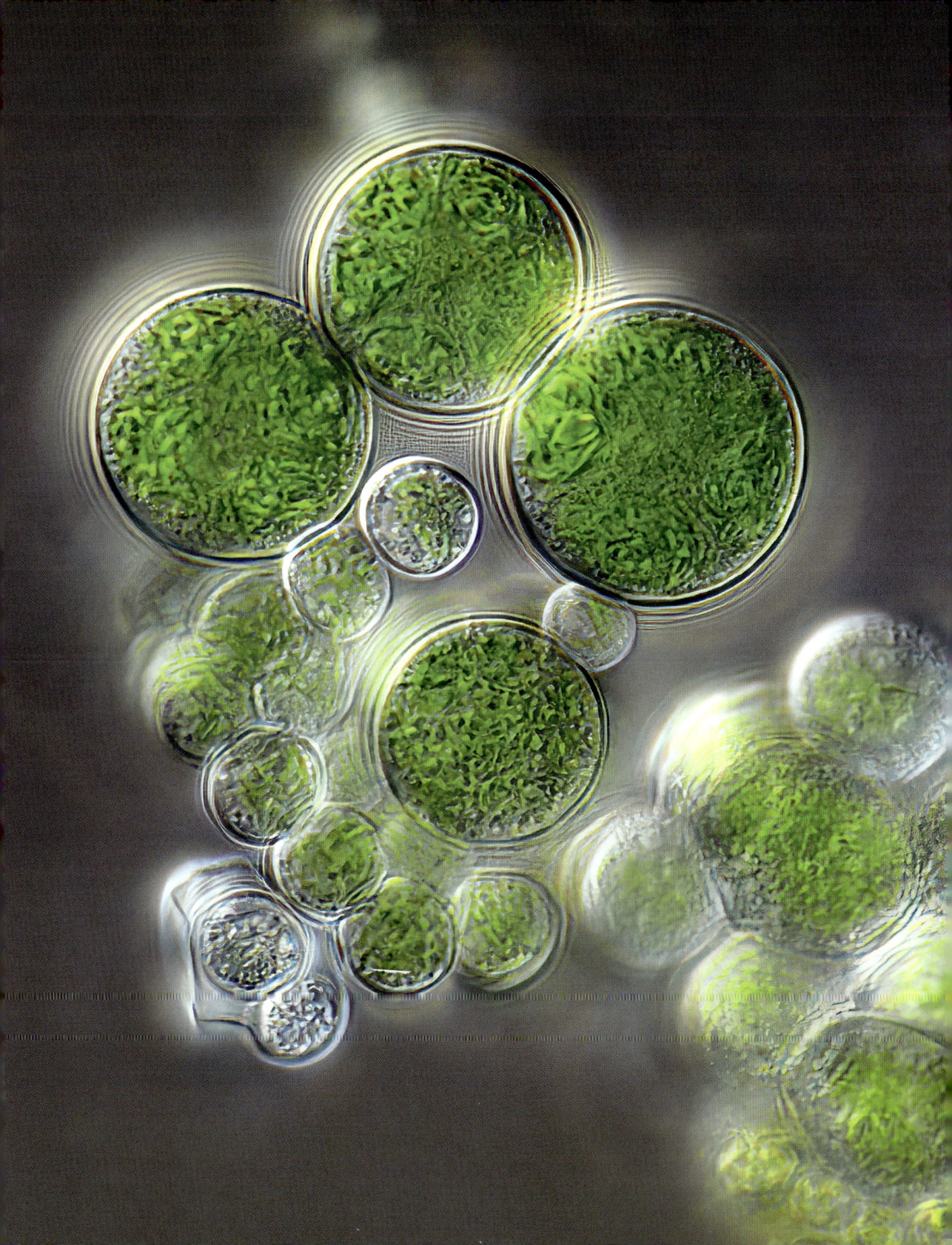

TRENTEPOHLIA AUREA

Golden Trentepohlia

Orange is the new green

SCIENTIFIC NAME	*Trentepohlia aurea* (L.) C. Martius
PHYLUM, FAMILY	Ulvophyceae, Trentepohliaceae
GROWTH FORM	Single to multiple chain-forming cells with one or more large carotenoid bodies, giving the alga a carrot-orange color
SPECIES IN GENUS	54
HABITAT	Common in foggy coastal regions on trees and rocks, and inland near the ground on tree trunks
NOTABLE FEATURES	A common lichen photobiont

Like the other major algal genus we profile here, *Trebouxia* (page 60), *Trentepohlia* is a genus of "aeroterrestrial" algae—in other words, algae that live on land, exposed to air. But in many ways, that is where the similarities end.

Unlike *Trebouxia*, *Trentepohlia* algae are capable of complex multicellularity. Together with a few close relatives, they are also unique among algae in possessing phragmoplasts, which students of plant anatomy might recall are cellular structures involved in laying down the beginnings of new cells.

Furthermore, despite being formally classified as green algae, they are usually orange in appearance on account of large carotenoid pigment bodies contained in every cell.

Trentepohlia algae are important symbiotic partners of fungi in lichens in several different distinct biomes. They feature prominently in tropical lichens that grow in deep undergrowth of rainforests, but also appear in high-humidity coastal environments from the Mediterranean to Baja California. Despite this decidedly beach-holiday ecology, species of *Trentepohlia* can also pop up as symbionts in some of the most inhospitable places on Earth, at the bases of trees close to the Arctic treeline. And just in case you thought you had the ecology of the genus narrowed down to humid environments, they also occur in lichens (and indeed also as free-living algae) on the sparse vegetation of the Atacama Desert in Chile, one of the driest places on Earth.

Approximately 20 μm

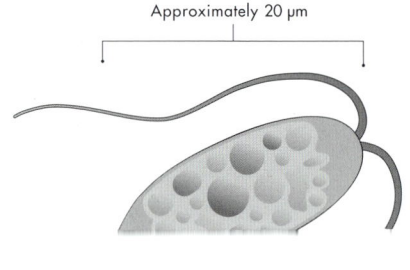

Mobile
The life cycle of *Trentepohlia aurea* includes flagellate spores. These are only known from free-living algae and not from the algae when in symbiosis.

→　*Trentepohlia aurea*, on rock in the mountains of Colombia.

Ule's Leaf-Dwelling Brownspot

Photobiont going rogue

SCIENTIFIC NAME	*Microtheliopsis uleana* Müll. Arg.
PHYLUM, FAMILY	Ascomycota, Microtheliopsidaceae
GROWTH FORM	Tiny crust lichen forming round, brownish patches
SPECIES IN GENUS	4
HABITAT	Strictly on living leaves of tropical rainforest plants
NOTABLE FEATURES	The photobiont always produces reproductive structures in this lichen

Given that lichens are known as slow-growing organisms, their occurrence on a short-lived substrate such as leaves is surprising. Leaf-dwelling lichens have solved this conundrum not by growing faster but by reaching maturity at extremely small sizes. One such lichen is *Microtheliopsis uleana*, happily reproducing with spores made in dark perithecia. And the photobiont joins in the reproductive frenzy.

The phyllosphere is a world of its own. It is a miniature ecosystem made up of organisms inhabiting the surface of living plant leaves. Given that the trees in temperate forests are typically leafless for several winter months, the phyllosphere is best developed in tropical rainforests. It consists of lichens and other surface fungi, liverworts and mosses, algae and cyanobacteria (and other bacteria), and invertebrates.

A common element of the phyllosphere is Ule's Leaf-dwelling Brownspot. It is found in rainforests on all continents. Its characteristic feature is the small, round, brown patches that develop spore-bearing, dark perithecia, often arranged in a concentric pattern. It also has what look like tiny hairs. These are actually the sporangia of the photobiont, an alga of the genus *Phycopeltis*, closely related to the aerial alga *Trentepohlia*. *Microtheliopsis* is the only genus of lichen fungi in which the photobiont produces such conspicuous reproductive structures with regularity. Why the alga reproduces in this manner in the presence of this particular fungus is unknown, but it suggests that they have a rather unusual relationship. Live and let live?

→ *Microtheliopsis uleana* growing on a leaf in a Costa Rican rainforest. The tiny hair-like structures are the sporangia of the associated alga.

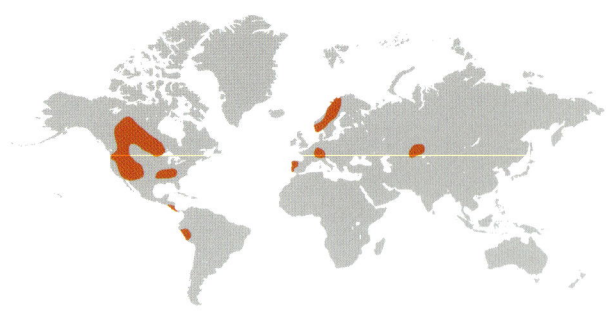

Lichenihabitans bacterium

Obligate lichen dweller

SCIENTIFIC NAME	*Lichenihabitans* spp.
PHYLUM, FAMILY	Alphaproteobacteria, Beijerinckiaceae
GROWTH FORM	Bacterial colonies, in lichens incorporated in and on the cortex
SPECIES IN GENUS	1, plus many unnamed species
HABITAT	Originally cultured from *Psoroma antarcticum* lichens in Antarctica
NOTABLE FEATURES	Represents a wider group of bacteria found worldwide on lichens

Most people, lichenologists included, do not think of bacterial cultures when they think of lichens. They were not one of the symbionts originally discovered in lichens, and until recently it was not even known that they were a constant part of lichen life. But like us humans, lichens have distinctive bacterial assemblages. The role they play in the lives of lichens is still being studied.

The culturing of bacteria from lichens began in the 1920s (page 54). As with exploration of the human gut microbiome, however, it was not until recently that technologies have been developed to survey the vast bacterial world associated with lichens using DNA sequencing technology. It turns out that many of the bacterial strains that have been isolated from lichens are one-off occurrences that may be picked up from the environment, but a few seem to have a strong association with lichens and are more or less always there.

Although it is early days in this line of research, this appears to be the case with the bacterial genus *Lichenihabitans*. The genus was given a name in 2019 by a Korean research group led by Yung-Mi Lee, based on a strain isolated from the lichen *Psoroma antarcticum*. Subsequent screening of DNA data from lichens has shown that close relatives, probably also members of the genus *Lichenihabitans*, are found in over 90 percent of lichens sampled from different parts of the world. What *Lichenihabitans* does in lichen symbiosis is not known with certainty. Genomes from closely related strains have been predicted to be able to synthesize vitamin B_{12}, which photobiont algae need to make the amino acid methionine.

→   A species of *Lichenihabitans* in culture at the University of Alberta.

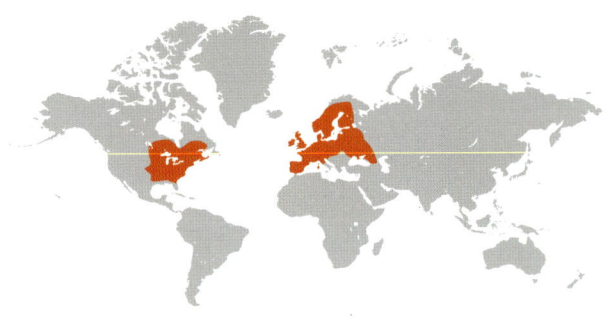

Marchand's Coral Fungus

Team player and killing machine

SCIENTIFIC NAME	*Marchandiomyces corallinus* (Roberge) Diederich & D. Hawksw.
PHYLUM, FAMILY	Basidiomycota, Corticiaceae
GROWTH FORM	Lichen-attacking fungus producing pink-red fruiting bodies
SPECIES IN GENUS	4
HABITAT	On a diversity of host lichens, but often on Parmeliaceae
NOTABLE FEATURES	Teaming up with another lichenicolous fungus to overcome host defense

We have seen that lichens are not just the stereotypical symbiosis between fungi and algae or cyanobacteria. Rather, lichens are miniature ecosystems involving and supporting a range of organisms. These also include a diversity of other lichen-inhabiting (lichenicolous) fungi. Not surprising, given that parasitism is one of the major fungal lifestyles.

Some of the lichenicolous fungi seem to live in peaceful coexistence with the main lichen fungus, taking advantage of the photobiont as commensals, but others fiercely attack the host.

The pink–red reproductive structures (bulbils) of *Marchandiomyces corallinus* are a beautiful sight. But make no mistake: this basidiomycete fungus kills. And it does so on a broad range of lichen hosts.

Studies led by lichen ecologist James Lawrey have shown that lichens do have chemical defenses against this fungus, one of these being lecanoric acid, another gyrophoric acid, the same lichen substances employed in dye production (page 250). These apparently "multitasking" compounds inhibit the growth of the attacking fungus and its cell-wall-degrading enzymes.

But then there is another fungus, the ascomycete *Fusarium*, which is able to degrade these lichen acids, overcoming the host defense. Once the *Fusarium* attacks a lichen, the *Marchandiomyces* (and other lichenicolous fungi) may come in and take advantage of the weakened host. Notably, *Fusarium* is the asexual morph of a species of *Nectria*, some of which are used as biological control agents. A striking example of all the things lichens can teach us.

→   The pink-red pustules (bulbils) of the lichen-attacking basidiomycete fungus *Marchandiomyces corallinus* make for a striking color splash. For the host lichen, here a species involving a fungus from the Parmeliaceae, *Parmelia sulcata*, the attack usually ends in death, at least for portions of the thallus.

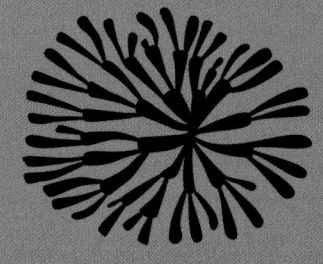

THE BIOLOGY
OF LICHENS

A day in the life of a lichen

In the Japanese anime film *Princess Mononoke*, forest spirits called *kodama* climb to the treetops at the first rays of light and watch the Night Walker morph back into the Forest Spirit. As the Night Walker appears, they rattle their heads, as if resetting themselves for the day. They are signs of a healthy forest, and their lives are centered around a narrow window of time in the daily cycle. Lichens operate on a similar cycle, typically centered on a short period around dawn.

PHOTOSYNTHESIZE, WAIT, RINSE, REPEAT

In the heat of the day, a lichen dries out. Its physiological activity trails off into the imperceptible, and it is in a state of semi-dormancy. But as night falls, temperatures drop and humidity rises, and it's time for the lichen to "wake up." The organisms that make up the symbiosis begin their respective routines. For the fungus, this might mean mending broken cells or adding new ones, or replenishing depleted secondary metabolites. Some algal cells might undergo cell division, making new cells. The previously dry thallus relaxes, unfurls, its symbionts respiring moistly and silently into the night.

But energy burned is energy that needs to be replenished. As the night draws to a close, and the sun rises, it's dinnertime for that hub of the lichen wheel, the photobiont. Though it has been active all night, it cannot photosynthesize without light. And now it has only a limited amount of time to do so. As the sun rises in the sky, temperatures warm and lichens quickly dry out again. This is the all-important window of time during which these biological *kodama* fix the carbon they need to replenish their daily stocks.

No sooner has this brief photosynthetic feast finished than the lichen is dry again. From now until the humidity rises again, the lichen waits, with very little, if any, measurable physiological activity in any symbiont. Innumerable variations on the tempo and pattern of this wetting–drying theme exist, depending where on Earth we are. In some habitats, lichens may not reach the required humidity to "wake up" and respire, let alone photosynthesize, every night; in others, they might be rocked out of bed midday by a rain shower: *wake up, dinnertime* all at once. In yet others, the dry nap might occur in just a narrow time window.

For most lichens, drying out doesn't seem to be optional; most species appear to *require* their daily dry nap. At a cellular level, the lichen undergoes a remarkable transformation. Cells shrivel, their organelles get squished together, and their once-watery cytoplasm goes into progressively drier states described as rubbery and, if taken to its desiccated extreme, glassy.

→ Awaiting its daily dose of hydration: the iconic lichen *Ochrolechia subplicans*, a denizen of coastal rocks in Alaska and northwestern Canada.

↓ Dry, Coal Miner's Snot Lichen.

WAIT, RINSE, REPEAT

Brittle and with somber colors, the Coal Miner's Snot Lichen (*Enchylium tenax*) looks rather unappealing when dry. However, like other jelly lichens, when wet it swells, soaking up water up to 500 percent of its dry weight. Jelly lichens owe this phenomenon to their photobiont, the cyanobacterium *Nostoc*.

In jelly lichens, the *Nostoc* colonies retain their copious gelatinous matrix, helping the lichen to retain water longer than in other lichens. *Enchylium tenax* is a component of biological soil crusts in arid regions. Given just enough moisture, these biocrusts form a thin "skin" over soil, helping to prevent erosion. The importance of biological soil crusts has consequently come to be recognized in land management everywhere, from Africa to the Gobi Desert to the American Southwest. That the *Nostoc* in these lichens also fixes nitrogen constitutes added value in ecosystems with limited nutrients.

↓   Wet, Coal Miner's Snot Lichen.

Temperature: 39.9 °F (4.4 °C)
Relative humidity: 83%

Temperature: 30.2 °F (−1.0 °C)
Relative humidity: 93%

Temperature: 54.0 °F (12.2 °C)
Relative humidity: 77%

Temperature: 57.7 °F (14.3 °C)
Relative humidity: 63%

This process, referred to in biology as vitrification—being turned into glass—is found in a smattering of different organisms, including resurrection ferns, mosses, and the larvae of some insects, which pass time waiting for the next dose of water. It is something we humans struggle to relate to, an alien superpower that enables organisms we would assume to be dead to spring back to life. How exactly lichen symbionts pull off this feat is not understood, but there are hints that these abilities may be the very thing that the symbiosis enables.

SUGAR ALCOHOLS: JANUS CARBOHYDRATES

During its photosynthetic frenzy, the photobiont fixes carbon and builds up an internal pool of carbohydrates. As briefly discussed on page 26, these are usually sugar alcohols, also known as polyols, and after they are transferred to the fungus, they tend to remain in this specific molecular form. This is not a given—many other forms of life would convert their carbs into other flavors, so the apparent insistence of lichens on keeping their carbs in the form of sugar alcohols has raised some scientists' eyebrows.

Sugar alcohols can be burned for energy, and most biology textbooks still report this as their sole function in lichens. But lichen researchers in the 1970s began to question that narrative, not least because most lichen

fungi grow exceedingly slowly for an organism that is ostensibly being bottle-fed with a nutrient subsidy. Indeed, sugar alcohols have other functions. Unlike non-alcoholic sugars, they are not highly chemically reactive, but like all alcohols they tend to extract and replace water.

Far away from the world of lichens, their properties fuel a multi-billion-dollar industry. Sugar alcohols are used as bulking agents in food, and in the biotechnology sector they are used to make sure that proteins survive desiccation. The basic idea behind the latter application is that proteins, fragile molecules that may be necessary as active ingredients in, say, a drug or vaccine, can be delivered to their final destination intact by being placed in a concentrated solution of sugar alcohols and then dehydrated, whereupon they become vitrified. In this line of business, sugar alcohols are referred to as excipients—essentially specialized delivery media.

NUTRIENT OR DESICCATION-SURVIVAL ELIXIR?

Sugar alcohols as well as other "non-reactive" sugars have indeed been found to be retained in very high concentrations in plants and animals that pull off the feat of vitrification (and the various stages of desiccation that lead up to it). Researchers in the 1970s began asking whether this was the case for lichens, too. Is it possible that lichen symbionts are not burning

Temperature: 59.2 °F (15.1 °C)
Relative humidity: 60%

Temperature: 59.0 °F (15.0 °C)
Relative humidity: 55%

Temperature: 58.1 °F (14.5 °C)
Relative humidity: 52%

Temperature: 50.0 °F (10.0 °C)
Relative humidity: 65%

↖↗   From green to gray: a time series showing a Freckled Pelt Lichen (a species in the *Peltigera aphthosa* group) drying out following a light rain.

the carbs fixed during photosynthesis primarily for growth, but instead are filling their cells with sugar alcohols to help shepherd their proteins (and DNA, and other molecules) intact through the hot, dry day until nightfall—as a kind of elixir enabling them to survive desiccation?

The possibility that fungi and algae maintain high concentrations of algal-derived sugar alcohols to get them through their dry nap raises all kinds of questions. In theory, the fungus could also burn its sugar alcohol pool for energy, but does it? If it does, how is overconsumption regulated? Can some lichens use algal-derived sugar alcohols, while others cannot? If a lichen fungus cannot burn the sugar alcohols, or only certain sugar alcohols, where does it get its carbon from?

OTHER POTENTIAL CARBON SOURCES

Fungi in lichens that grow on trees might get at least some of their carbon the way other fungi do—by breaking down plant-derived carbohydrates like cellulose. Perhaps tellingly, analyses of genomes have shown that most lichen fungi still have the potential to make the prerequisite enzymes to convert these carbohydrates into food. However, many lichens grow in places like rocky deserts where sources of carbon are not exactly plentiful. What other carbon sources might be on the table?

Perhaps the most obvious alternative source of carbon—other than the direct products of photosynthesis—is the cell walls of dying algae, but how quickly that carbon would be taken up depends on the life span of the algal cells within the lichen, which nobody knows. Another possibility is that fungi take up carbon through the spongy lichen cortex layer, which soaks up water and everything dissolved in it. Some back-of-the-envelope calculations have suggested that the amount of carbon found in dust and rainwater might just be sufficient to cover fungal growth. It is also possible that different lichens use different mixtures of carbon sources. Tellingly, most lichens exhibit very slow growth compared to other fungi or to plants, which means that they probably do not need much carbon overall.

A FUNGAL MARSHMALLOW TEST?

In the early 1970s, psychologists began testing short-term versus long-term gratification in people. A child was offered one small, but immediate, reward or two rewards if they waited. The go-to rewards in early tests were marshmallows. If the marshmallow was left untouched after 15 minutes, the child could have two of them. Children who waited for the second marshmallow tended to have better life outcomes (probably they are our bosses today).

The dual uses of sugar alcohols present lichen symbionts with a similar dichotomy. If they burn them, they can grow, and probably grow fast; other fungi certainly burn sugar alcohols and grow just fine. This is the immediate, one-marshmallow scenario. Alternatively, if they do not burn them—save the marshmallow— they can use them to survive progressively longer and dryer desiccation episodes, and thus gain access to a habitat few fungi otherwise do well in: the harsh, sunbaked environments exposed to the open air.

Perhaps the most tantalizing aspect of a lichen symbiosis model under which the alga provides the fungus with sugar alcohols to get them through their daily drying routine is that it gives a possible explanation for the incredibly slow growth of lichen fungi. Lichen fungi almost certainly evolved slow growth from faster-growing ancestors, which means, in the barebones world of natural selection, that faster-growing individuals were eliminated from successive populations. This is a remarkable paradox, considering that lichen fungi are almost universally classified as recipients of nutritional subsidies. The possibility that the very subsidies popularly thought of as nutritional could have dual use (or possibly even be walled off for a *non-nutritional* use) sets up a mechanism by which carbon-greedy fast-growing individuals would have been knocked out of the population during extreme wetting–drying events.

→ The Palmetto Lichen (*Ramalina celastri*) has been used to study the metabolism of lichens in relation to desiccation and rehydration, and the role that sugar alcohols play in it.

Slow and old

Most of us old enough to remember 30 years back might have a frame of reference for the passing of time in the living world. Perhaps a niece or nephew was born and has since gone on to have children of their own. Perhaps a seed you picked out of an apple core and planted is now a tree that shades your garden and produces more fruit than anyone can eat. Or perhaps, like some people, you have gone back to the same rock year after year to monitor how a lichen has grown a healthy 5/16 in (8 mm) around its edges.

Such are the lives of lichens. Nothing happens too hastily. We know this because, as long as there has been lichenology, there have been people who want to know how fast these strange things grow, and what kinds of past events can be reconstructed by knowing how old they are. This pursuit is called lichenometry— the measuring of lichens.

OPERATING ON LICHEN TIME

The use of lichens to date past events was pioneered in the Austrian Alps in the 1950s, to assess the speed of glacier retreat by measuring thallus sizes of lichens closer to and farther from the glacier edge. The new science of lichenometry was quickly adopted by geologists, and the method also came to be applied in other arenas, including the dating of earthquake faults, rockfalls, debris flows, and snow events.

Lichenometry has also been used in archaeology to date important historical events. In one of its first and more famous applications, German lichenologists attempted to reconstruct the age of the famous moai statues of Easter Island, concluding that most of the statues were erected around the year 1530.

Lichenometric calibration is also the source of the widely repeated claims of the great age of lichens, including estimates of more than 11,000 years for Map Lichens from the Brooks Range of northern Alaska.

Not everyone is convinced. Severe criticism has been leveled at the methodology behind lichenometric dating. One of the weaknesses in the method is the assumption that the largest thalli are those that colonized soon after the substrate became available, and that they persist indefinitely. Detractors argue that thallus mortality is common, even within observable time frames. They also criticize the transfer of calibrated "growth curves" generated in one study to other, independent, studies.

→ *Buellia frigida*, the "frigid" Icecold Button Lichen, lives up to its name. For most of its time it just lies around doing nothing. It is found in the Dry Valleys of continental Antarctica, one of the harshest environments on Earth. With growth rates as low as 0.0036 mm per year, about the thickness of an individual fungal hypha, individuals of this lichen are estimated to be at least 6,500 years old, meaning they would have witnessed the end of the Stone Age. (Inset shows a close-up.)

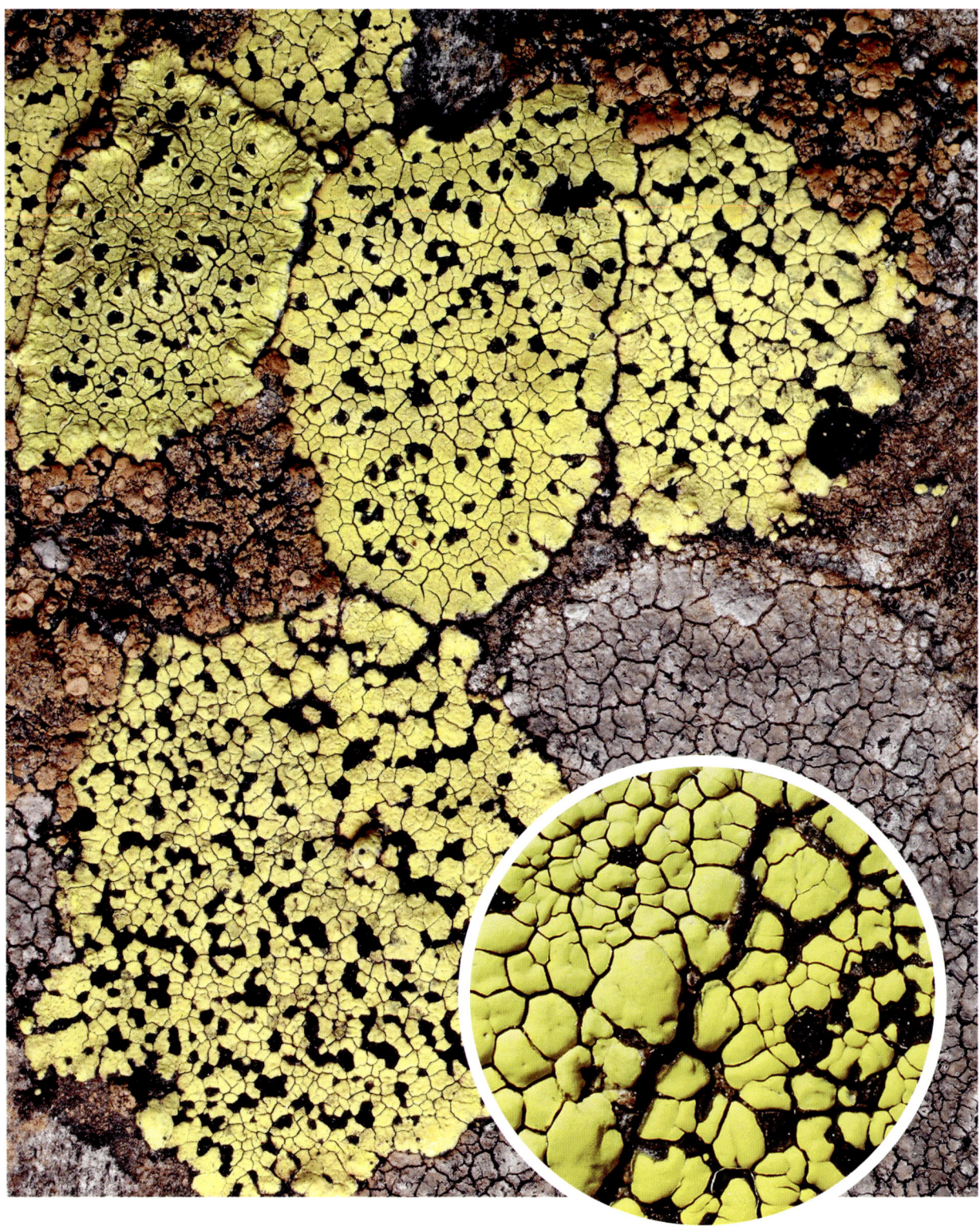

Methodological reservations about lichenometry notwithstanding, the slow growth rates of lichen fungi, relative to their non-lichen relatives, are not disputed. Lichen fungal cultures have been estimated to grow at one nine-thousandth the rate of the lab model fungus *Neurospora crassa*, and even the fastest growth rates measured in nature are a tiny fraction of the growth rates of other fungi, or plants for that matter. Why is this?

Scientists have found evidence that slow growth is hardwired into lichen fungal genomes, specifically into the genetic code for translating DNA into proteins. This might explain why lichen fungi grow slowly even when they are provided with excess nutrients, and it may be an evolutionary consequence of the penalization of fast use of resources (page 78).

Some of the record holders for fast-growing lichens are hair lichens of the genus *Usnea* and the Lace Lichen (*Ramalina menziesii*). In these lichens, growth rate is compared in terms of the percent thallus size increase over a set time period, and a few *Usnea* lichens, such as the Methuselah's Beard Lichen (*U. longissima*), have been recorded as doubling their thallus length in under a year. It is unknown what allows some lichens to exhibit such growth while most do not.

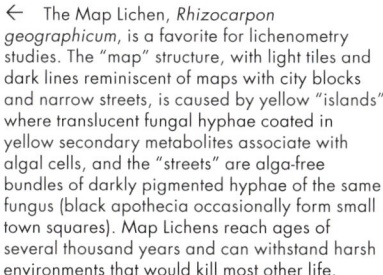

← The Map Lichen, *Rhizocarpon geographicum*, is a favorite for lichenometry studies. The "map" structure, with light tiles and dark lines reminiscent of maps with city blocks and narrow streets, is caused by yellow "islands" where translucent fungal hyphae coated in yellow secondary metabolites associate with algal cells, and the "streets" are alga-free bundles of darkly pigmented hyphae of the same fungus (black apothecia occasionally form small town squares). Map Lichens reach ages of several thousand years and can withstand harsh environments that would kill most other life.

→ The pros and cons of lichenometry. The slow growth of this *Rhizocarpon* lichen was documented and measured through a photographic time series (A–D). However, at some point, the lichen died off and left an empty space, allowing for succession long after initial colonization of the bare rock surface.

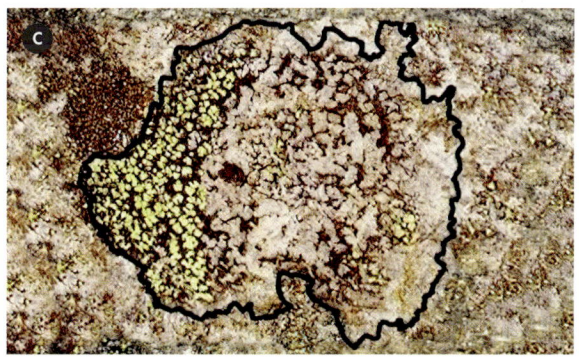

A protective patina

Many of the lichens familiar to the nature-loving public are colorful—yellow splotches on garden walls and neighborhood trees, red lipstick lichens, purple Christmas wreaths. All of this color is chemistry on display.

When in symbiosis, the interactions between fungi and algae give rise to molecules called secondary metabolites. They are called secondary because they are not essential to the life-sustaining processes of primary metabolism, such as respiration and cell replication. But for the lichen, secondary metabolites may define where it grows, what light the alga receives, and whether it has a place in the ecosystem's food chain.

Crystalline deposits of secondary metabolites can be seen under the microscope and usually make up between 1 and 5 percent of the dry weight of a lichen or, in certain specialists, as much as 30 percent. Such is the case for the lichen *Ingaderia friabillima*, which lives on cacti and sieves water vapor out of fog in the Atacama Desert. At last count over 1,000 different secondary metabolites have been characterized and named by chemists.

IN A SPECTRUM OF THEIR OWN

Some of these substances are not visible at all in wavelengths visible to us humans (about 400–700 nm) but fluoresce in ultraviolet light, meaning that they convert the invisible UV light into visible light of different colors. Ultraviolet light can be used to bring out these colors, the most striking one being the bright yellow caused by a substance called lichexanthone. Some animals—including birds, lizards, and insects—can see light in the UV range, but it is not known what they see when they look at a lichen that contains these substances.

Secondary metabolites in the UV spectrum are more abundant in lichen groups that are diverse in lower temperate and tropical regions of the world than in the cooler upper temperate and subpolar regions.

This makes for fun activities when strolling in a subtropical or tropical forest at night with a UV flashlight, the lichen-covered tree bark lighting up in diverse colors.

CHEMICAL DEFENSES

Dedicating up to 5 percent—or more—of your body weight to one type of molecule is costly, and it suggests that secondary metabolites are important to lichens. Different theories have been advanced, and it is possible that no single explanation applies to all. Many substances may even be multifunctional.

What is clear, however, is that many secondary metabolites deter grazing. "Lichen grazing" takes many forms. Caribou or Reindeer (*Rangifer tarandus*) live off of lichens during the winter months. Microscopic animals live *in* them and make meals of the lichen cortex and the starch-rich fruiting bodies, the apothecia. Important among these smaller lichenivores are barklice and mites. In humid forests around the world, snails and slugs can strip the lichen cortex and algal layers clean, leaving behind white zigzag "grazing marks" where only fungal hyphae remain visible.

Another group of animals that feed on lichens are oribatid mites. Over 80 species have so far been documented in lichens, but the real number is likely much higher. Very few experimental studies (maybe none) have been conducted on secondary metabolites and mites, but anecdotal evidence suggests that the red pigment rhodocladonic acid, which gives the crimson red in lipstick lichens, is formed in response to mites burrowing and feeding on the fungal mycelium (page 28).

← Lichen rave? A tree trunk in the National Botanic Garden in Havana, Cuba. At night, illuminated by a UV flashlight, the otherwise mostly white lichens light up in a spectacular mosaic of different fluorescent colors.

↑ *Pyxine* lichens (here the widespread *Pyxine cocoes*, the Coconut Palm Rosette Lichen) are literally the "stars" among fluorescent lichens, with their bright yellow reaction caused by the chemical substance lichexanthone, a common "sunscreen" in tropical lichens (upper two images). The related genus *Dirinaria* went down another path: whereas the thallus absorbs the UV light completely, the soredia fluoresce in light blue, caused by the secondary metabolite divaricatic acid (lower two images).

→ Many lichen communities, particularly on rock in exposed situations, are composed of brightly pigmented species, often dominated by yellow Acarosporaceae and yellow-orange to orange-red Teloschistaceae forming a color mosaic.

PROTECTION FROM THE SUN

Many types of secondary metabolites, typically located in the upper part or cortex of the lichen, function as natural sunscreens. Experiments carried out both directly on lichens and on purified extracts of lichen secondary metabolites have shown that these substances filter light in the UV-A part of the spectrum, within the range that can be carcinogenic for humans. Obvious examples of such lichen sunscreens are the UV-fluorescent substances mentioned above, but other common ones include yellow to red anthraquinones and brown to almost black melanins. Their production in response to sunlight can be observed in the field: lichens of the same species growing in more exposed sites will have more of the pigment than those growing in more shaded conditions. In full sunlight, the Yellow Wall Lichen (*Xanthoria parietina*) is bright yellow to almost orange, but in the shade it is gray with only yellow margins.

Some of the secondary metabolites found in lichens have been studied for possible pharmaceutical applications (page 242). The advantage for the lichen, however, likely lies less in the prevention of cancer, which is unknown from fungi, rather than in the filtering of light to the alga. The photosynthetic apparatus in algae can be damaged by UV-B radiation (280–315 nm) and generally by intense light, but this is filtered out by lichen sunscreens, especially when the lichen is dry. When lichens are wet, the cortex swells and more light passes through; as seen from the outside, the effect is that the lichen appears greener. This phenomenon has been called the "cortical window."

EVOLVED TO BE INDESTRUCTIBLE

Lichens grow in some of the most extreme environments on our planet, including in places where little else thrives. Name the most extreme terrestrial habitats on Earth, and more likely than not there's a lichen that specializes in making a living in them.

At least 25 species of lichens live on rocks in the dry valleys of Antarctica, one of the coldest and driest places on Earth, with temperatures regularly dipping to −60 °F (−51 °C) and precipitation a meager 2 in (50 mm) per year. They have been found at elevations higher than that of the highest vascular plants or mosses in the Himalayas. They coat the rocks of the Atacama and Namib deserts, where their thalli thrive on surfaces hotter than 160 °F (71 °C), around the temperature at which eggs begin to fry. Lichens obtain the little water they need to live here from early-morning fog from the nearby oceans, but rarely see a drop of rain.

Experimental extremes

As if these feats in the natural world were not enough, lichens have been subjected to even greater extremes in the name of science. Some of these experiments sound like something between candidates for an Ig Nobel Prize and entries in the *Guinness Book of World Records*. A dry sample of the Yellow Wall Lichen (page 36) has survived immersion in liquid nitrogen at temperatures below −295 °F (−182 °C), not to mention being bombarded with gold and palladium particles and examined in a vacuum chamber with a focused electron beam. The algal symbiont of a lichen has been brought back to life after several decades stored dry in a museum collection, where it had been fumigated with the pesticide bromomethane.

Given all that lichens can handle, it is only logical that they would attract the interest of astrobiologists, scientists interested in the potential of life outside of our own planet. On May 31, 2005, lichens were sent into space aboard a Soyuz rocket launched from the Baikonur Cosmodrome in Kazakhstan. They spent 15 days on the outer panel of a Russian satellite as part of a European Space Agency (ESA) experiment called BIOPAN-5. The lichens survived exposure to the full ultraviolet light spectrum of space, unfiltered by our atmosphere; gamma radiation; and vacuum conditions. Between 70 and 90 percent of the lichens on this mission continued to exhibit "vital signs" after return to Earth.

Numerous subsequent experiments have been undertaken on the ability of lichens to weather space conditions. In another ESA research project called Expose-E, samples of the Map Lichen (page 82) were externally mounted on experimental panels of the International Space Station (ISS) for 18 months between 2008 and 2009. Astonishingly, 10–15 percent of lichen samples survived. Lichens were exposed for a similar amount of time during a 2014–2016 stint aboard ISS panels. Yet more studies have been performed in facilities that simulate space or Mars-like conditions.

Staying dry is key

All of these feats are only possible if a lichen is dry. Wet the lichen, and all bets are off. The rare ability of lichen cells to spring back to life after lying dormant in a vitrified, glass-like state (page 72) seems to be the ticket to successfully negotiating all manner of natural and unnatural mischief inflicted upon them. Whatever Hollywood may try to spin, just remember that being able to vitrify like a lichen might just be the best recipe yet for surviving the apocalypse.

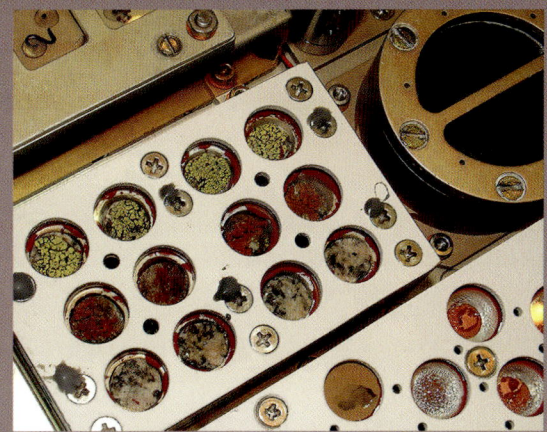

↖↗ The Biology and Mars Experiment (BIOMEX), containing lichens, was hosted on the EXPOSE-R2 panel on the International Space Station for 469 days between 2014 and 2016, with temperature variation between −4° F and +136° F (−20 °C to +58 °C) and high gamma radiation. Photos show the setup in space and its preparation and the capsule after landing.

Environmental monitoring and protection

Lichens have evolved to withstand some of the harshest conditions found in terrestrial ecosystems, in part owing to their capacity to handle prolonged desiccation. Notwithstanding this toughness, many lichen symbioses are sensitive to environmental change, including pollution, altered land-use, and a warming climate.

Even in the surrounding country it was a foggy day, but there the fog was grey, whereas in London it was, at about the boundary line, dark yellow, and a little within it brown, and then browner, and then browner, until at the heart of the City—which call Saint Mary Axe—it was rusty-black.

Charles Dickens, *Our Mutual Friend* (1865)

The Industrial Revolution brought infrastructure and mechanization but also heavy air pollution. And it was not only the characters in Charles Dickens's stories that were suffering. The first recorded decline of urban lichens goes back to 1859, perhaps unsurprisingly in the English city of Manchester. It did not take long for lichenologists elsewhere in Europe to see the same phenomenon in their own cities. From the 1860s onwards lichen decline was reported in Paris, London, Newcastle-upon-Tyne, Glasgow, Augsburg, and Munich; by the early 1900s lichen decline was being reported from all across the continent.

AIR FIT FOR LICHENS

Realizing the potential of this grim state of affairs to inform public policy, researchers in several European countries and the United States began to experiment with using lichens as a biological indicator of air quality and, by extension, an early warning system for human health. In the 1920s, the Swedish lichenologist Greta Sernander devised a classification system whereby city districts could be mapped into three broad categories: "lichen desert," "struggle zone," and "normal zone."

At a time when automated equipment for measuring air quality in terms of particulate matter and pollutants such as nitrous oxides and sulfur dioxide was not yet widely available, lichens offered a cheap and easily assessed metric for air quality. Today, lichens are widely recognized as environmental indicators by the general public. "Air fit for lichens, water fit for trout" was an early environmental rallying cry attributed to the ecologist Kenneth Mellanby.

The underlying reasons for lichen decline were long debated, but few lichenologists seemed to doubt that suspended chemical pollution played a role. Some pointed to greater dieback associated with humidity, and available evidence does suggest that poor air quality is compounded by humidity, likely because these are the times when lichens are physiologically active. Research in the 1970s and 1980s confirmed that sulfur dioxide gas leads to the degradation of chlorophyll in the alga (and also in plants in general), suggesting that the photobiont is the "weak link" in maintaining lichen symbioses in this specific kind of pollution.

ACID RAIN AND EUTROPHICATION

The industrial age has been a bleak time for lichens.
No sooner had the rusty-black fog of Victorian London
cleared than acid rain began falling across large parts of
the industrialized world. This phenomenon peaked in
the early part of the Cold War, and precipitated not
only the mass dieback of forests but the widespread
decline of lichens. Here, too, sulfur dioxide has been
implicated as exercising a toxic effect on algal
photosynthesis.

Acid rain was eventually overcome by the
implementation of industrial filters and catalytic
converters, but fast on its heels came another wave of
bad news for lichens in the form of a massive increase
in the use of industrial fertilizers in agriculture. The
deposition of aerosolized nitrogen and phosphorus has
reached proportions unprecedented in Earth's history,
a phenomenon called eutrophication.

Different lichen symbioses exhibit profoundly
different responses to eutrophication, with a few, such
as the Yellow Wall Lichen, benefiting, while most others
disappear. Robust data are hard to come by, and it can
be challenging to disentangle whether past declines
were due to fertilizer drift or sulfur dioxide deposition.
Regardless, the cumulative toll on lichens has been
severe. In the German state of Lower Saxony, for
instance, about 40 percent of the lichens historically
found in the state have not been seen since the end
of the nineteenth century. This dieback has hit certain
types of symbioses particularly hard, including the
so-called cyanolichens, in which fungi are twinned
with cyanobacteria.

→   When first described in 1883 from Epping Forest
in northeast London, England, *Lecanora conizaeoides*,
the Dusty Rim Lichen, was characterized as "the only
well-developed species" in an otherwise diminished lichen
community of the area. Only much later was it realized
that this lichen seemed to thrive under atmospheric
pollution, and it became a signature lichen for polluted city
areas. Curiously, with improved air quality, it has strongly
declined and is now considered rare.

BEARING ELEMENTAL WITNESS

Sometimes, lichens don't disappear, but rather quietly soak up the secret results of human mischief. Lichens were among the organisms to register elevated levels of radioactive caesium–137 in the North American Arctic following Soviet and Chinese nuclear weapons tests in the 1970s, though they did not appear to be any the worse for wear on account of it. Biomonitoring approaches such as the United States Forest Service Air Quality Monitoring Program have capitalized on the propensity of lichens (and the mosses with which they often cohabit) to bioaccumulate elements, to develop massive geographical data sets across public lands that reveal subtle shifts not only in composition but also in elemental deposition, including in the form of elements that might not immediately be harmful to the lichen itself. These kinds of data are used to inform government regulators.

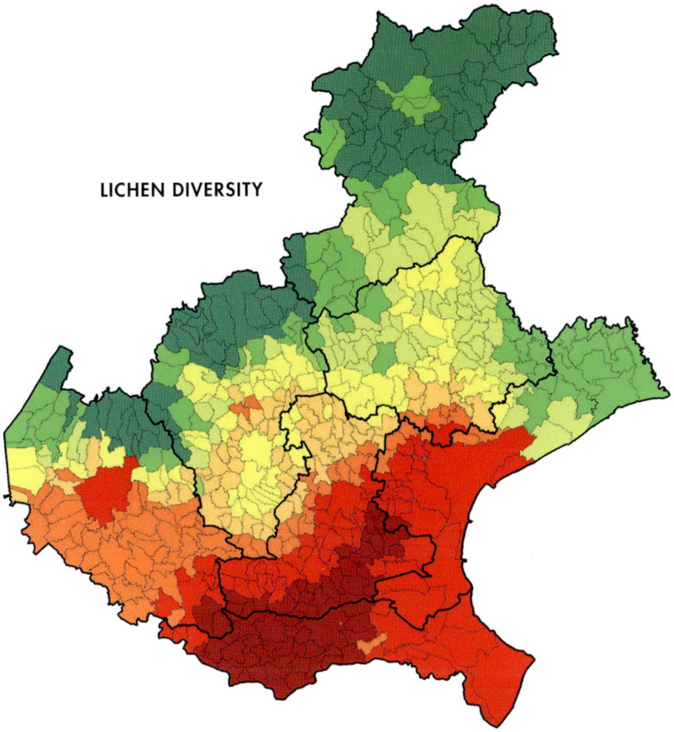

LICHEN DIVERSITY

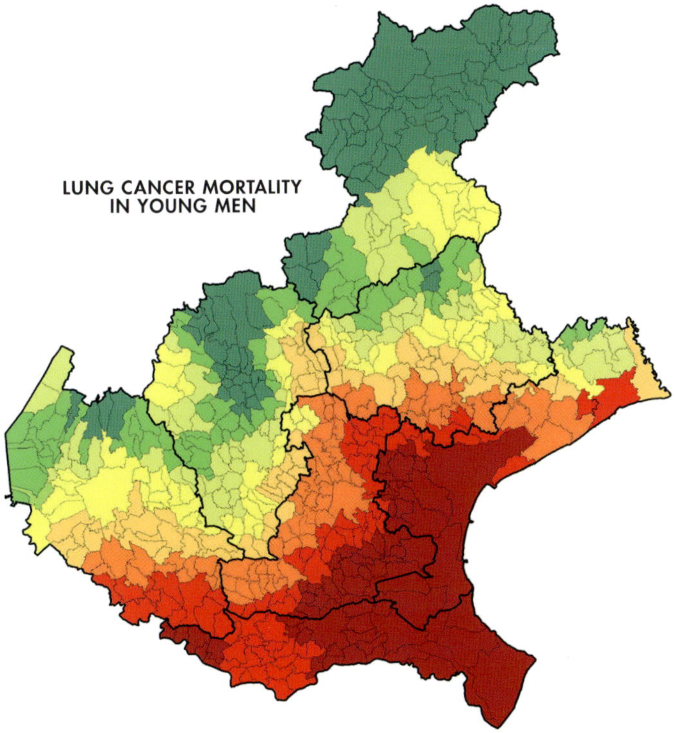

LUNG CANCER MORTALITY
IN YOUNG MEN

Air bad for lichens, bad for people

In 1997, Italian lichenologist Pier Luigi Nimis and medical statistician Cesare Cislaghi reported on striking correlations between lichen diversity and lung cancer incidence in young men in northern Italy, both correlated with high levels of atmospheric pollutants, such as ozone (O_3), sulfur and nitrogen oxides (SO_x, NO_x), carbon monoxide (CO), and volatile organic compounds (VOCs). In these maps, showing the provinces within the region of Veneto, the color gradient goes from green (high lichen diversity and low lung cancer incidence), through yellow to red (low lichen diversity and high lung cancer incidence).

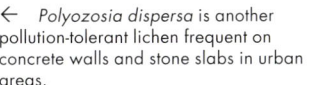

←   *Polyozosia dispersa* is another pollution-tolerant lichen frequent on concrete walls and stone slabs in urban areas.

Lichens in a changing climate

We've seen that lichens are sensitive to environmental changes they did not evolve to deal with. They are not adapted to acid rain and cannot escape these kinds of adverse conditions by going dormant. Lichens get soaked with excessive nutrients or other chemicals, and most disappear from the landscape. However, there is one major challenge for all of life on Earth that may end up being the biggest of all: climate change. What does a changing climate mean for lichens?

Although lichens are generally assumed to be outstandingly sensitive to changing environments, relatively few studies have been undertaken to test how they will respond to climate change. Perhaps because of their remarkable resilience against natural environmental extremes when dry, the assumptions applied to other organisms, such as death by heat, might not apply. However, other types of conditions may be lethal.

A promising approach to understanding lichen response to climate change is to induce it experimentally. This is what has been done in a forest experiment called SPRUCE in northern Minnesota, in the United States. Here, in a 20-acre (8 ha) Black Spruce (*Picea mariana*) peat bog, plexiglass enclosures have been erected in which temperature can be raised by up to 16 °F (9 °C) compared to surrounding ambient temperatures and carbon dioxide concentrations can be doubled. The idea is to create conditions that scientists predict will exist in our children's lifetimes. And here, Abigail Meyer, Daniel Stanton, and other members of their team from the University of Minnesota have been monitoring what happens to the lichen *Evernia mesomorpha*.

RISING TEMPERATURES INDUCE THALLUS BLEACHING

Meyer and her team have discovered that carbon dioxide levels have little detectable effect on two key lichen performance metrics, photosynthesis and carbon respiration. However, rising temperatures prove lethal. Any increase beyond 2 °C (3.6 °F), expected to be reached globally by the middle of the century at current rates of warming, results in the death of the algal symbiont, followed after a lag time by severe degradation of the fungus. The result is thallus bleaching: the thalli lose their natural color and become gradually whiter and whiter. Eventually they die and fall to the ground.

It is not clear what kills the alga, but Meyer and Stanton have a hypothesis. Much as humans notice changes in their metabolism at the same temperature depending on whether it is dry or humid (think of the discomfort of "mugginess"), organisms that photosynthesize react to the differences in air moisture content, and warm air can hold more water than cooler air. This greater capacity means that warm air pulls moisture out of lichens much faster than cool air. Faster drying means less time for photosynthesis. As we have pointed out, the lichen can survive desiccation *per se*, but the "dinnertime window" in its daily wetting-drying cycle (page 76) becomes shorter and shorter.

↑→   The SPRUCE forest experiment
in northern Minnesota allows testing of
the effects of climate change on lichens
and other organisms.

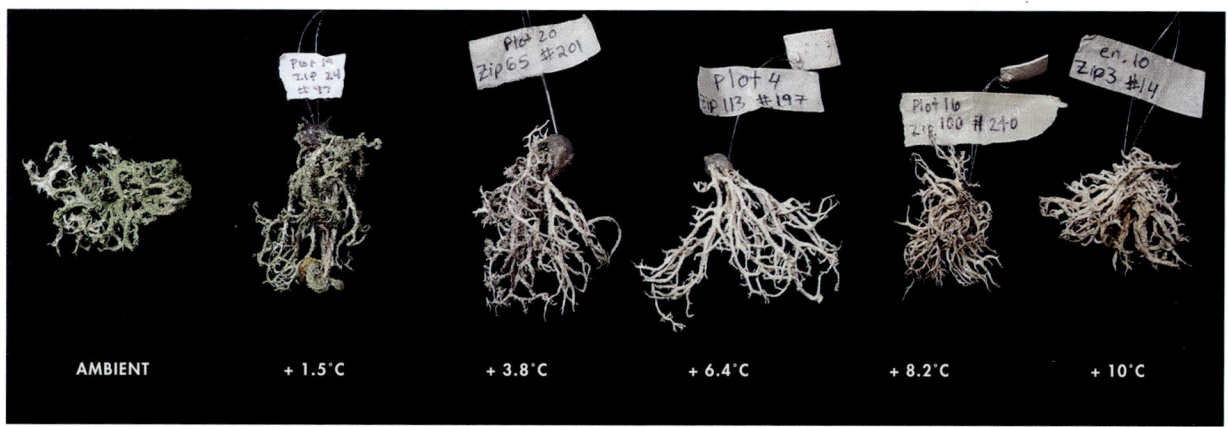

AMBIENT   + 1.5˚C   + 3.8˚C   + 6.4˚C   + 8.2˚C   + 10˚C

←↑ Lichen bleaching with experimentally increased temperatures, from the SPRUCE climate change experimental facility in northern Minnesota, USA, using the Boreal Oak Lichen (*Evernia mesomorpha*) as study object. The healthy lichen in its natural habitat is shown to the left.

This drying effect is made worse by another temperature effect. Although the rate of respiration (carbon loss) increases as temperatures go up, photosynthesis peaks—often between 59 °F and 77 °F (15–25 °C)—and then starts decreasing. So even as metabolic costs go up, the rates of carbon gain, and the time available to make it up, go down. This has far-reaching consequences for the amount of carbon dioxide that can be turned into sugar alcohols every day.

LESSONS FROM CORAL REEFS?

What Meyer and her team have found in the Minnesota climate enclosures has echoes in the coral bleaching that has started to become more frequent owing to warming oceans in the last 20 years. Coral bleaching is also caused by the death of a photosynthesizing symbiont (zooflagellates, which are also often called "algae"), and the coral animal has a finite amount of time to acquire new symbionts or recover before it, too, dies. Researchers are looking at the feasibility of supplying corals with new, warmth-tolerant algae to extend the life of the reef.

So far too little is known about lichen bleaching to predict whether experience gained with coral reefs will help lichens. The Minnesota team found that before they died completely, lichens in early bleaching stages exhibited higher levels of algal genetic diversity than healthy lichens, suggesting that the ailing lichen was harboring a wider range of algal strains as the symbiosis broke down.

Whatever the environmental conditions that lead to dieback, scientists have little doubt that lichens will respond—in one way or another—to climate change. Those that can move fast enough might migrate latitudinally, or upwards to higher elevations. Others might ultimately disappear if there is no place else to go. But will anyone notice? We'll need a much higher density of observations than we have at present, and many dedicated observers returning to the same trees and rocks year after year, to determine real changes in species composition. This is a promising future area of collaboration between lichenologists and citizen scientists.

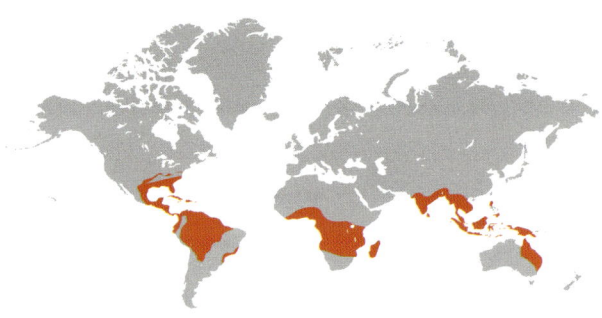

HAEMATOMMA ACCOLENS

Tree Bloodspot

Vivid color for UV protection

SCIENTIFIC NAME	*Haematomma accolens* (Stirt.) Hillmann
PHYLUM, FAMILY	Ascomycota, Haematommataceae
GROWTH FORM	Small crustose lichen with disk-shaped apothecia
SPECIES IN GENUS	50
HABITAT	Tree bark and fence posts in tropical to subtropical woodlands
NOTABLE FEATURES	Readily recognized by the red disks with white margins

Bloodspot lichens live up to their name: the fruiting bodies with their strikingly red disks can be spotted from a distance, and so these lichens are commonly collected and photographed. The precise function of the red pigments is not known, but it can be assumed that they serve to protect the underlying hymenium where the ascospores are formed, since red pigments absorb potentially damaging UV radiation.

The genus *Haematomma* is unique among lichens and easily recognized: no other lichens combine a crustose, white thallus with the fruiting bodies (apothecia) featuring a red disk bordered by a white margin. The long, narrowly fusiform ascospores with multiple septa are also characteristic.

The distinction of species is more complicated, and depends on checking for subtle differences in the composition of secondary chemical compounds that can only be identified with sophisticated methods such as chromatography. A small portion of the lichen, finely ground, is treated with acetone, which extracts the substances. The extract is then spotted onto a glass plate covered with finely ground silica gel. When the plate is placed in a tank with an organic solvent, the substances move toward the top of the plate, and how far they reach in a given time depends on how hydrophilic ("liking" water) or hydrophobic ("hating" water) they are. Their final position and their spot characteristics (like color) are then used for their identification.

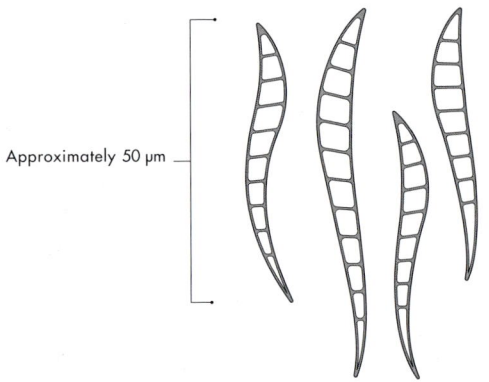

Approximately 50 μm

Means of dispersal

Haematomma accolens reproduces by means of characteristic spores (ascospores) produced in the bright-red fruiting bodies (apothecia). Each spore consists of several cells arranged in a straight row.

→   Thallus of *Haematomma accolens* photographed in a tropical rainforest in southeastern Brazil. After rainfall, the lichen becomes fully hydrated and ready for photosynthetic action.

STELLARANGIA ELEGANTISSIMA

Showy Namib Firedot

Fractal patterns under the desert sun

SCIENTIFIC NAME	*Stellarangia elegantissima* (Nyl.) Frödén et al.
PHYLUM, FAMILY	Ascomycota, Teloschistaceae
GROWTH FORM	Crust lichen with closely attached, radiating lobes
SPECIES IN GENUS	4
HABITAT	Coastal deserts in southern Africa (Namib Desert)
NOTABLE FEATURES	The lobes sometimes grow erect, away from the substrate

A desert is not the sort of place you would expect to find lichens. And yet, coastal deserts in South America, southern Africa, and western Australia harbor some of the most spectacular lichen assemblages known on this planet, characterized by brightly colored representatives of the Teloschistaceae. The genus *Stellarangia* is perhaps the most striking of these, turning pebbles into small jewels.

The evolution of a similar body architecture under strong selective pressure is an evolutionary phenomenon found not only in lichens, but it is particularly obvious in these symbiotic organisms, and notorious in the family Teloschistaceae. Indeed, more than a dozen genera in this family, often only distantly related, look almost exactly the same as the lichen depicted here, although few are as spectacular.

Firedot lichens used to be classified in a single genus, *Caloplaca*, encompassing all crustose species in the family Teloschistaceae. Other than being crustose, *Caloplaca* species are very diverse in appearance and therefore were long considered an artificial grouping. DNA sequence data confirmed that *Caloplaca* is not a consistent unit, but also showed that lichens that looked similar were not necessarily closely related. The most striking case is that of the placodioid species, now classified into more than a dozen genera. One of these, *Stellarangia*, is endemic to coastal deserts in southern Africa. It has a similar-looking cousin in the South American Atacama Desert, the genus *Follmannia*.

→   The spectacular *Stellarangia elegantissima* adorns Namibian rocks and pebbles.

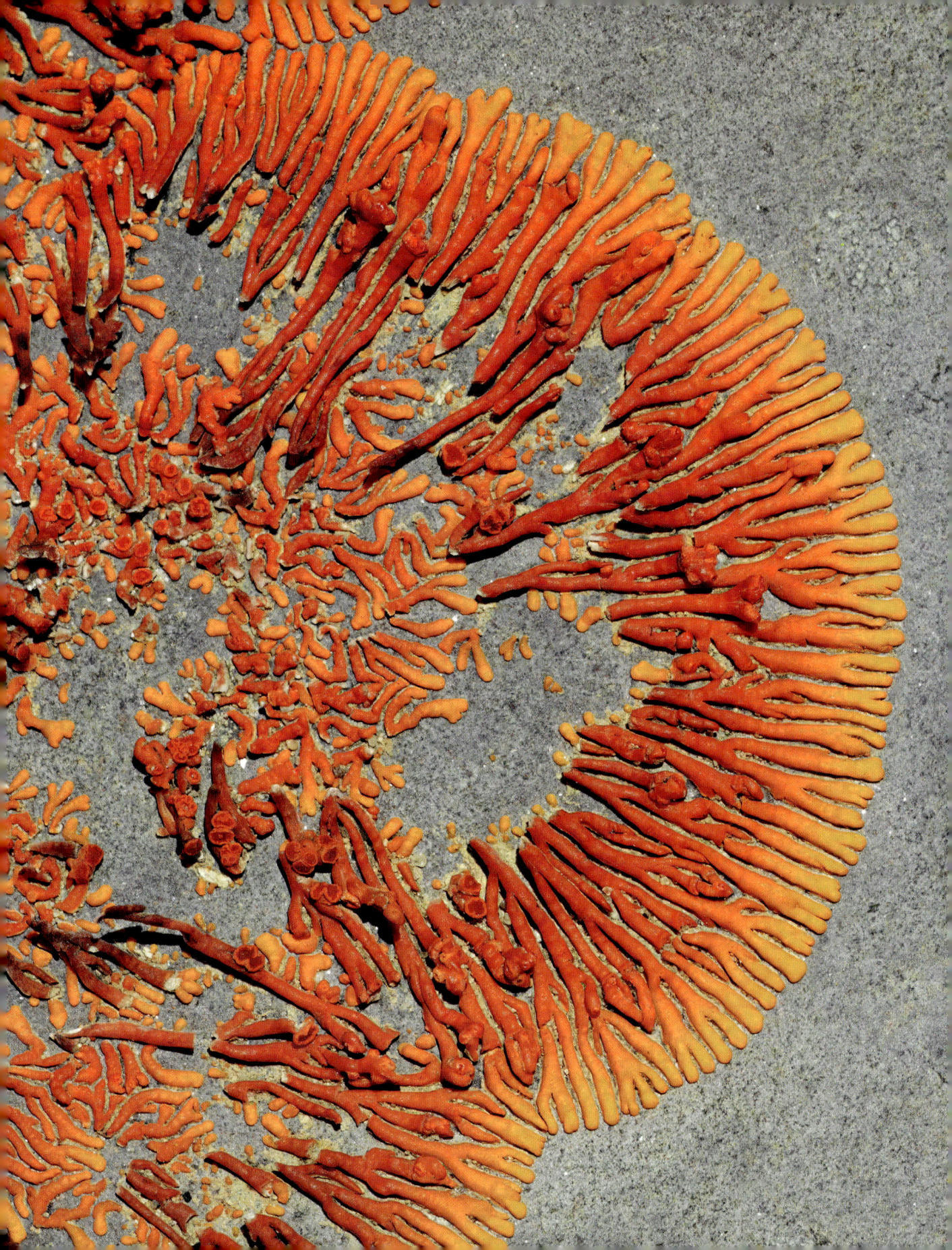

DERMATISCUM THUNBERGII

African Gold Lichen

Coating cliffs in southern Africa

SCIENTIFIC NAME	*Dermatiscum thunbergii* (Ach.) Nyl.
PHYLUM, FAMILY	Ascomycota, Caliciaceae
GROWTH FORM	Large umbilicate foliose lichen that is canary-yellow above and brown-black on the underside
SPECIES IN GENUS	3
HABITAT	On sheer rock faces
NOTABLE FEATURES	One of the few umbilicate members of its family

Among the most easily recognized lichen growth forms are the so-called "umbilicate" lichens, in which the thallus is shaped like a saucer and attached to its substrate at a single point in the middle, reminiscent of an umbilical cord. This type of growth form has evolved independently multiple times, almost always on rock.

Large vertical cliffs are special habitats for many species of lichens. Why large cliffs support species not found on small cliffs is not obvious, but some species clearly specialize in this habitat. Perhaps they require a specific humidity regime that guarantees constant ventilation. Or perhaps they require the massive thermal updrafts associated with large vertical cliffs for their dispersal.

In southern Africa, *Dermatiscum thunbergii* is one such specialist. Its upper surface is loaded with the secondary metabolite rhizocarpic acid, the same substance that lends the yellow color to the Map Lichen (page 82). Its hues paint entire cliff faces. Both its chemistry and its growth form are highly unusual for the group of lichen fungi it is related to, the Caliciaceae, which mostly include crust and pin lichens, whereas umbilicate lichens typically involve fungi from the aptly named Umbilicariaceae.

→   African Gold Lichen can cover cliff faces on such a large scale that it can be seen, and positively identified, from miles away.

Elegant Sunburst Lichen

Experimental model of astrobiology

SCIENTIFIC NAME	*Rusavskia elegans* (Link) S. Y. Kondr. & Kärnefelt
PHYLUM, FAMILY	Ascomycota, Teloschistaceae
GROWTH FORM	A foliose lichen with neatly radiating orange lobes
SPECIES IN GENUS	18
HABITAT	On highly basic rocks like limestone, and on concrete
NOTABLE FEATURES	Survived a trip to space without a space suit

Lichens that grow in sun-exposed habitats are often brightly colored on account of their sunscreen pigments, and the Elegant Sunburst Lichen is no exception. It is one of the lichens most commonly photographed by hikers on the rugged limestones of the Alps and Rocky Mountains, but can be found in suitable habitats from the Arctic to the Antarctic.

Sunscreen pigments come in various forms, including bright yellow and wavelengths outside our visible spectrum. A large number of lichens, including the Elegant Sunburst, have settled on various shades of orange. These are usually anthraquinones, a group of substances with various industrial applications that are widespread in nature, including in places as diverse as rhubarb root and aloe.

It was the possession of its specific orange sunscreen that led to the Elegant Sunburst Lichen being chosen as one of the model lichens by researchers working on the Expose-E project on the International Space Station in 2008 (page 88). Pieces of live *Rusavskia elegans* and other lichens were launched into space aboard the Space Shuttle *Atlantis* and brought back to Earth aboard the *Discovery*—and *R. elegans* was among the "best survivors." Like many lichens, its ability to shut down until better times come along, behind a veil of secondary metabolite crystals, gives it abilities that are out of this world.

5 mm

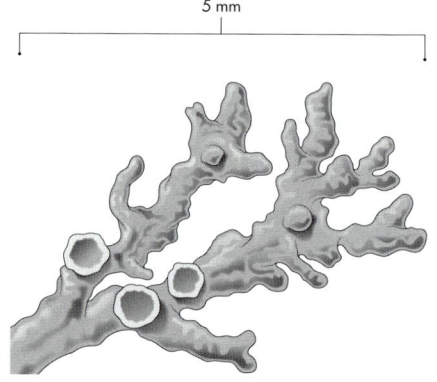

Zooming in
Detail of the lobes of *Rusavskia elegans* bearing apothecia.

→ *Rusavskia elegans* forming one of its characteristic elegant sunbursts, here on a canal wall in the middle of the city of Berlin, Germany. The thallus divides into outwardly radiating lobes, while the fruiting bodies, the apothecia, form in the older, central parts of the thallus.

Hooded Tube Lichen

Ready-to-go packets under the hood

SCIENTIFIC NAME	*Hypogymnia physodes* (L.) Nyl.
PHYLUM, FAMILY	Ascomycota, Parmeliaceae
GROWTH FORM	A foliose lichen with hollow lobes and open, hood-like lobe tips
SPECIES IN GENUS	About 90
HABITAT	On trees, shrubs, and fence rails, sometimes also on vertical rock faces
NOTABLE FEATURES	An extremely common and adaptable lichen found around the world

The Hooded Tube Lichen has all you need to know to identify it packed into its name: no other common lichen has both a hollow thallus and lobe tips that open up into soredium-lined hoods.

The Hooded Tube Lichen is one of the commonest lichens on trees in the northern hemisphere. It is part of a broader group of lichens involving fungi from the family Parmeliaceae. These so-called "parmelioid lichens" form many of the familiar broad-lobed foliose species with differently colored upper and lower surfaces. The majority of lichens, whether they grow on rocks or trees, are found mainly in nutrient-poor places. Historically, this would have given them many places to live—most natural woodlands are nutrient-limited, with the available nitrogen and phosphorus largely spoken for by the trees themselves.

With the felling of forests for agricultural lands in places like eastern China, Europe, and the American Midwest, woodlands have increasingly become part of a patchwork quilt mixed with heavily fertilized crop agriculture. This has led to a big decline of parmelioid lichens.

Although the Hooded Tube Lichen is considered a pollution-sensitive species in areas with severe air pollution, it is one of the first parmelioid lichens to reappear when conditions improve. Why this is the case is not known, but one reason could be its photobiont. The Hooded Tube Lichen is a symbiosis of the fungus *Hypogymnia physodes* and the relatively uncommon alga *Trebouxia suecica*.

→ The Hooded Tube Lichen (*Hypogymnia physodes*) prefers somewhat acidic substrates but even so is sensitive to air pollution.

LICHEN
ARCHITECTURE

Fungi in the service of photosynthesis

Lichens are the only fungal symbioses in which form has evolved in response to photosynthesis. Neither of the main symbionts can take on the form on its own; the architecture of each lichen develops only when the fungus is in the presence of its photosynthetic partner. We are only beginning to understand why the forms we see today prevailed, but each three-dimensional lichen outcome almost certainly represents a sweet spot on a spectrum that integrates optimal exposure to light, on the one hand, and optimal thallus ventilation, on the other.

BEAUTIFUL BY NATURE

At the beginning of the twentieth century, the naturalist and artist Ernst Haeckel published a compilation of lithographs titled *Kunstformen der Natur* ("The art forms of nature"). Among the 100 prints, which range from single-celled organisms to flowering plants and vertebrates, plate 83 was dedicated to lichens. Haeckel described these as "one of the most curious classes of the plant kingdom." He acknowledged their symbiotic nature, emphasizing that both the external shape and the internal anatomy are driven by the intimate consortium of algal cells and fungal hyphae, unique among known life forms. He was particularly fascinated by *Cladonia* lichens and their relatives. The *Cladonia* growth form continues to be one of the most widely recognized lichen architectures, with their basal scales and erect, often trumpet-shaped podetia.

To this day, Haeckel's lithographs, reproduced as posters, adorn dorm room walls and the virtual walls of social media. Unbeknownst to many, the beauty Haeckel depicted contrasted with his harmful advocacy of racism and eugenics. Can the art truly be separated from the artist?

→ Haeckel's lichen plate from 1904, showing the Lung Lichen (*Lobaria pulmonaria*) in the center, surrounded by various species of *Cladonia*, *Pulchrocladia retipora* (above middle), and highly stylized thalli of *Anaptychia* (lower right), *Flavoparmelia* (lower left), *Melanohalea* (upper right), and *Physcia* (upper left). (See overleaf for a twenty-first-century equivalent.)

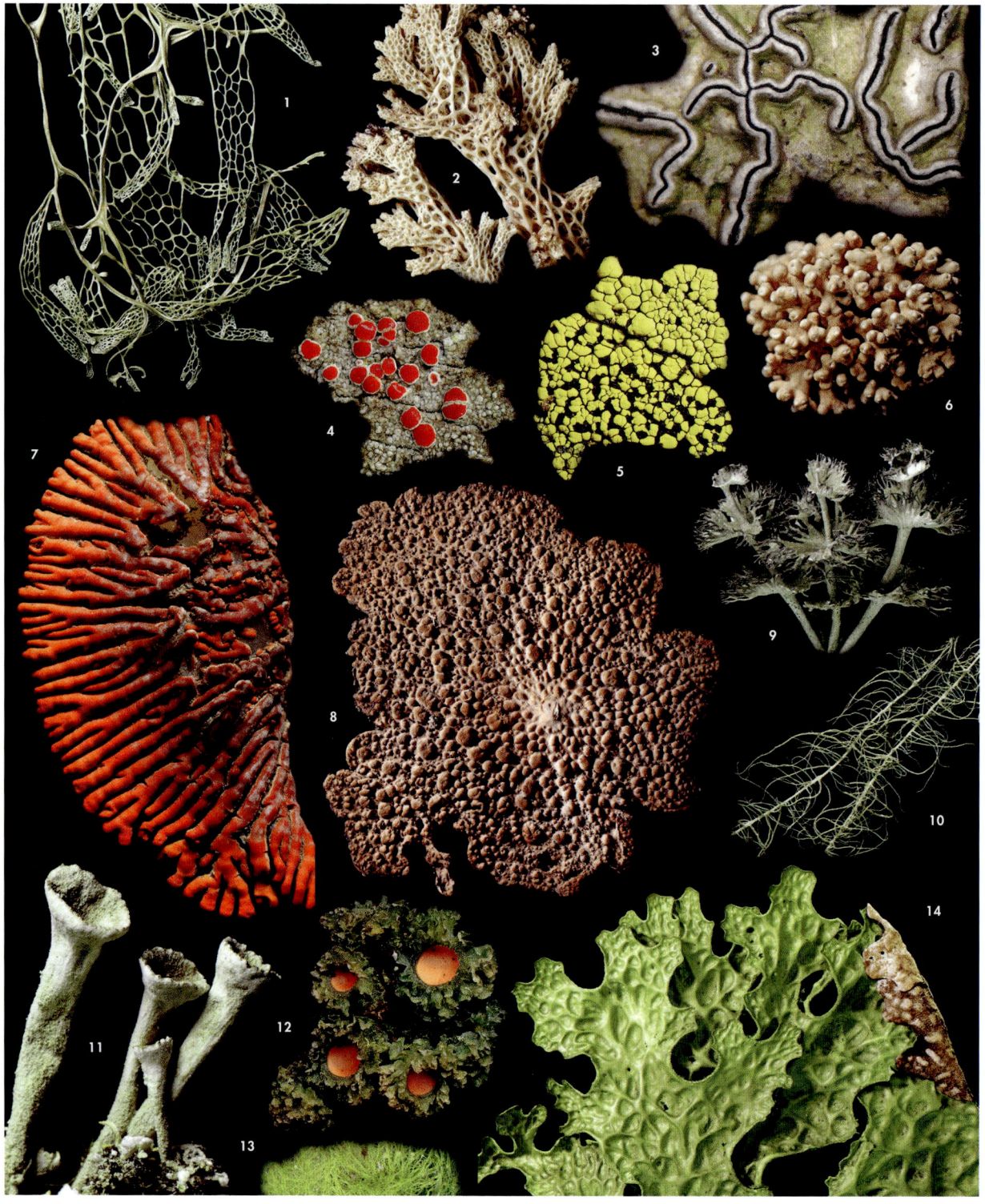

ARCHITECTURE DRIVEN BY SYMBIOSIS

In most lichens, the thallus—the "body" of the lichen—consists mostly of the fungus, which provides a structural scaffold for its architecture. Algae or cyanobacteria occupy a layer within. Exceptions are thread-like (filamentous) and gelatinous (jelly) lichens, in which the thallus appears to be shaped by the photosynthetic partner.

↑ *Rhizocarpon geographicum* growing on a piece of glass offers a unique view of the delicate architecture of this lichen.

← An attempt to bring Haeckel's idea (see page 111) into the twenty-first century using photographs. The modern plate depicts the main architectures, including crustose lichens (3, 4, 5, 7), foliose lichens (14), umbilicate lichens (8), jelly lichens (12), shrub and beard lichens (1, 2, 6, 9, 10), dimorphic lichens with trumpet-shaped podetia (11), and filamentous lichens (13).

The multicellular structure built by the lichen symbionts is not like those of animals. Instead of filling space by cell divisions, lichens can be thought of as analogous to wicker garden chairs in which individual strands of fungal hyphae, taken together, form a structure. As in the garden chair, a specifically shaped outcome is the result of thread-like fungal hyphae adhering to each other at specific angles. But the fungus does not do this in the absence of a photosynthetic partner, nor has a lichen fungus ever been observed to sexually reproduce in the absence of the latter.

It is still not understood what the alga or cyanobacterium triggers for the fungus to begin the cell-glued-to-cell self-organization that ultimately makes a lichen. The photosynthetic partner, by permitting the better fungal performers of this routine to reproduce more often, may hold the key to an age-old cycle of natural selection that gives us the diversity of lichen shapes we know today.

A MIRROR OF PLANT MORPHOLOGY?

Most plants adhere to one basic scheme: roots, stem, leaves. Lichens, by contrast, do not exhibit a single body plan and take on a variety of shapes, or growth forms.

The basic growth form of a lichen can be determined by answering a few simple questions:

- Is the growth two- or three-dimensional?
- Is the lichen attached (either firmly or loosely) to the surface, or does it have a single attachment point?
- Do erect structures emerge from a basal part of a different structure?
- Is the fungus the dominant element, or is the shape determined by the photobiont?
- Does the lichen swell when wet?

Fruticose
Lacking a clear upper and lower surface; can be beard-like, hair-like, or shrub-like (*Ramalina*)

Foliose
Leaf-like, with a well-defined upper and lower surface (*Coccocarpia*)

Crustose
So closely attached to its substrate that the lichen cannot be removed without destroying it; crust-like (*Graphis*)

Gelatinous
Like foliose but swelling and jelly-like (*Enchylium*)

Squamulose
Like foliose but with small discontiguous scales (*Flakea*)

Umbilicate
Like foliose but with central attachment point (*Umbilicaria*)

Filamentous
Like fruticose, but shape determined by algal or cyanobacterial threads (*Dictyonema, Coenogonium*)

Dimorphic
With erect structures (podetia) emerging from basal crustose or squamulose thallus (*Dibaeis*)

LEAF

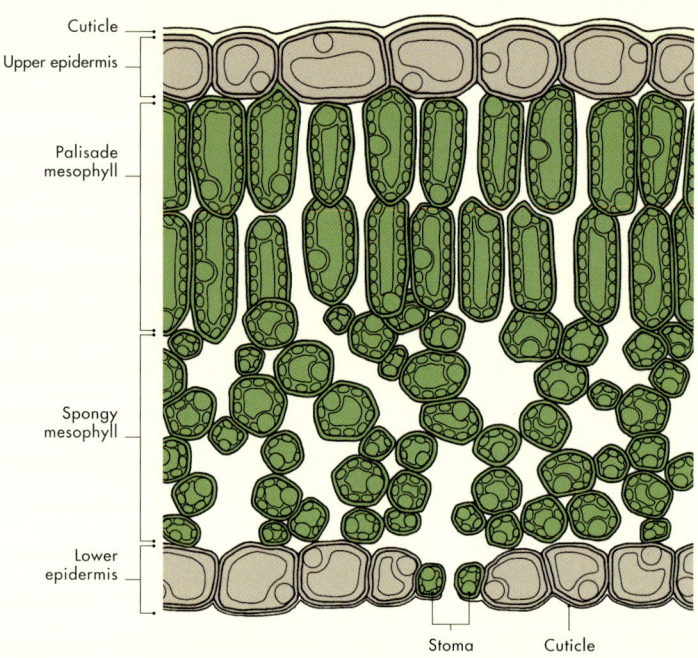

Cuticle
Upper epidermis
Palisade mesophyll
Spongy mesophyll
Lower epidermis
Stoma   Cuticle

LICHEN

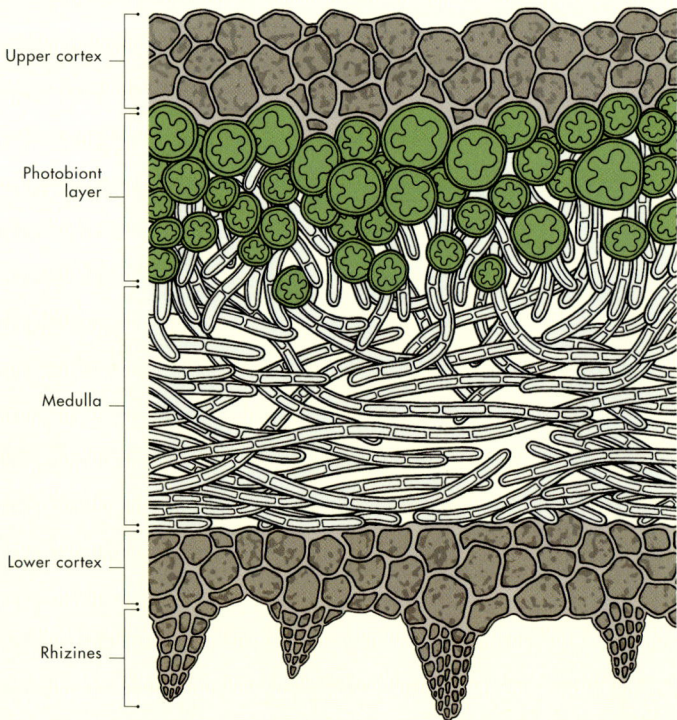

Upper cortex
Photobiont layer
Medulla
Lower cortex
Rhizines

Lichens and plants independently converged on remarkably similar solutions to some environmental challenges, such as the optimization of photosynthesis. A section through a layered lichen resembles a section through a leaf, although in some details they are also quite different. In many other respects, lichens and plants went different ways. Lichens take up water and nutrients regularly over their entire surface, while vascular plants (tracheophytes), including seed plants, ferns, and fern-like plants, use roots and specialized water-transporting vessels (although some uptake and loss of water and nutrients through leaf surfaces also occurs). Furthermore, almost all lichens are tolerant of desiccation at the cellular level. Only a few tracheophytes can do this, one example being the "resurrection plants" of the genus *Selaginella* found in many desert regions.

Conversely, some plants have converged on three-dimensional architectures remarkably similar to those of lichens in their evolution of habitat specialization. Bromeliads of the genus *Tillandsia*, relatives of the pineapple, resemble shrub or beard lichens and take up water and nutrients through the leaf surface, through special scaly hairs. These "atmospheric" bromeliads compete with lichens in situations where other plants would have difficulties, growing on branches and even telephone wires, side by side with shrubby to pendant *Ramalina* or *Usnea* lichens.

One of the most striking examples of plant–lichen convergence is *Tillandsia usneoides* ("resembling *Usnea*",) known as Spanish Moss across the southeastern United States (see photographs opposite).

↖← **Functional convergence?**
A section through a leaf compared with a section through a layered lichen thallus.

←↓ Plant or lichen? Spanish Moss (*Tillandsia usneoides*), not a moss but a flowering plant (left), is often mistaken for a beard lichen of the genera *Usnea* or *Ramalina* (below). This *R. anceps* was photographed in the Galapagos Islands.

FUNGAL HOMOLOGIES?

The similarities between lichens and plants are based on analogy—they did not acquire similar-looking structures from a common ancestor. The comparison between lichens and other fungi is another matter, based on homology. Lichens draw most of their biomass from interwoven fungal hyphae, the same kind of microscopic cellular filaments that make up mushrooms and molds. In fungi that are not involved in lichens—mushrooms, bracket fungi, and cup fungi—architecturally complex structures are dedicated to reproduction. But precisely how lichens achieve their architectural complexity is not really understood and has been described as "the holy grail of plant–microbe interactions."

↑   Similar hyphae, different outcome. In mushrooms, the vegetative body, the mycelium, forms a network of hyphae living in or on the substrate (above). In lichens, by contrast—here, the imposing *Cladonia imperialis* from Brazil—the entire mycelium is engaged in a complex, above-ground thallus architecture (above right).

In mushrooms, genes associated with sex are upregulated during the early stages of fruiting-body formation. One theory is that the lichen thallus, too, is derived from modified fruiting bodies. In the 1960s, the French lichen anatomist Marie-Agnès Letrouit-Galinou showed the striking similarity between the bundling of fungal hyphae in *Cladonia* podetia and hyphal bundling in non-lichen fungi such as *Heyderia*, which forms its spores atop small clubs. With the recent discovery that some club fungi such as earth tongues (*Geoglossum*) and some lichen fungi share a relatively close common ancestor, the plot thickens. The results of more studies are eagerly awaited.

Forms shaped by selection

Evolution by natural selection is a ruthless process. For any organism, what we see today is the outcome of everything else that was tried having been less fit—less successful, in other words, at passing on its genes to the next generation. In lichens, this means that if we see a lineage of foliose lichens, those lichens are foliose because any predecessors that took on a different form—perhaps as crust lichens—were less fit in the habitats in which this lineage occurs, and eventually disappeared from the gene pool. So how does this work?

[handwritten margin notes: "Obviously wrong", "predecessor in what sense?", "Surely not ancestor? But then what?"]

WHAT DRIVES SELECTION FOR THALLUS SHAPE?

As we saw at the beginning of this chapter, lichen fungi cannot make a lichen on their own, but when fungus and alga do interact to give rise to a lichen, it is the fungus that provides the scaffold through hyphal arrangement. Let's return to the wicker-furniture analogy, in which the fungus is the weaving material that makes up the bulk of the furniture (willow or rattan, for instance). But, as if by magic, the fungal weaving material is alive, and self-assembles when it senses the presence of algal or cyanobacterial cells, which sit like tennis balls within the furniture. Moreover, as we saw earlier, the fungus is only allowed to have sex when those balls are present, alive, and well.

Imagine a large population of such wicker-furniture-building fungi. Now enter natural selection: given specific environmental conditions, such as a tree branch in a rainforest, only those fungal individuals that build furniture shapes that allow the alga to be present, alive, and well will be able to reproduce. Those that do a lousy job will not, or will have fewer offspring.

Whatever that ideal shape is, the fungal individuals that achieve it will have a fitness advantage; those that offer slightly different shapes will carry fitness costs, produce fewer offspring, and eventually be removed by natural selection. What we see is the resulting thallus.

So what are some of the fitness costs that have shaped lichen thalli? One example that is comparatively well studied, and relevant to our rainforest tree branch, is sensitivity to saturation. We know most lichens can survive virtually anything dry, but when they are saturated, all bets are off. More specifically, most lichens—perhaps all—are far more sensitive to heat when wet than when dry, and many of the extreme feats they pull off when dry would not be possible when wet. Because of this, lichen evolutionary biologists have started to suspect that in environments with plentiful moisture, being unable to rid oneself of water is costly.

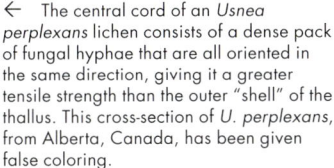

← The central cord of an *Usnea perplexans* lichen consists of a dense pack of fungal hyphae that are all oriented in the same direction, giving it a greater tensile strength than the outer "shell" of the thallus. This cross-section of *U. perplexans*, from Alberta, Canada, has been given false coloring.

→ The thalli of *Lecanora sphaerospora* (top) and *Psora crystallifera* (bottom) lichens, at home in coastal deserts of southern Africa and western Australia, resemble quartz crystals, with internal anatomies analogous to those of the popular houseplants known as "living stones" in the genus *Lithops*. The algae are stacked between vertical columns of fungal hyphae interspersed with large blocks of calcium oxalate crystals, an arrangement that appears to optimize the flow of light while reducing evaporation.

SELECTION FOR THALLUS SURFACE AREA

The foliose growth form affords lichens a convenient way to get air underneath the thallus, and to keep it away from other waterlogged objects, such as bark or moss. The Canadian lichenologist Trevor Goward has even argued that rhizines, the root-like appendages on the undersurfaces of countless foliose lichens, are not for attachment, as has widely been claimed, but are the result of selection for flood-zone stilts—pegs to help ensure that the thallus has just the underside ventilation it needs. No data have yet been gathered to test this hypothesis, but it checks many of the evolutionary theory boxes—bulky thalli without some means of ventilation are more likely to succumb to waterlogging. Perhaps tellingly, lichens in very wet habitats often go all in on maximizing surface area, with thin thalli, long stilts or cilia, and even hair-like morphologies. In all of these cases, the alternatives must have been more costly.

Does this mean that all lichens with expanded surface areas evolved this trait to shed water? Not necessarily. The evolutionary cost–benefit equation needs to be assessed in the context of the environment in which they live. Thallus shapes with maximized surface areas have also arisen in some of the driest habitats on Earth. In the Atacama Desert in northern Chile, entire years pass with almost no measurable rain, yet lichen thalli with highly increased surface areas are common. Here, it is possible that such thallus forms have been selected to sieve water from fog rolling in from the nearby ocean, like filter feeders. Ridding themselves of this water is not a problem, but getting too little would also carry a cost.

↗   Water from fog and clouds condenses as a droplet on the cilia of this montane rainforest lichen, *Leucodermia lutescens* (left). Highly dissected shrub lichens such as *Cladia aggregata* (right) exhibit the same phenomenon.

→   Leaf lichens (here *Sticta aongstroemii* from southeastern Brazil) can reach impressive sizes under favorable conditions, allowing them to compete with plants, such as ferns. Their plant-like shape also optimizes ventilation in these highly humid conditions.

OTHER FACTORS THAT DRIVE SELECTION

The costs associated with too much or too little water, and how these may have shaped lichen thallus surface area, are but one set of examples of the pressures that might have led to the lichen forms we see today. Other factors, such as light, likely shaped thallus forms, sculpting them over evolutionary time into natural solar panels. Nutrient scarcity in extremely nutrient-poor habitats, or the need to get rid of excess nutrients, such as in bird colonies, are doubtless additional selective filters. Our understanding of the evolution of lichen thallus form is in its early days, and lichen evolutionary biology is a promising field for more study.

→ Body plan is just one aspect shaped by selection. Color is another, and lichens are as colorful as they come. Lichens in light-exposed habitats often coat their thallus or their fruiting bodies with a layer of pigments providing protection against UV radiation. In the unmistakable *Ramboldia russula*, here photographed in North Carolina, the fruiting bodies are bright red.

CONVERGENCE ON THE SAME MOTIF
....

The same environmental stress that acts upon selection also filters particular architectures, leading to similar body plans in unrelated but cohabiting organisms, a phenomenon called convergence. *Candelina mexicana* and *Wetmoreana variegata* appear strikingly similar. Both grow on exposed rocks, and the yellow to orange pigments, though chemically different, protect against UV radiation. The fungi involved in these lichens diverged from a common ancestor about 300 million years ago.

Jelly lichens of the genera *Gabura* (here *G. fascicularis*) and *Collema* (here *C. curtisporum*) also demonstrate convergent evolution. Some of these have been classified as each other's closest relatives since 1780, and only recently have DNA data revealed that they belong to entirely different lineages.

↓ A. *Candelina mexicana;*
B. *Wetmoreana variegata;*
C. *Gabura fascicularis;*
D. *Collema leptaleum.*

CONVERGENCE AND DIVERGENCE

The ascolichen *Peltigera* (*P. canina*) and the basidiolichen *Cora* (*C. paraciferrii*) evolved convergent architecture and ecology through natural selection. The insides of these lichens reveal anatomical differences, demonstrating that their similar architecture did not evolve from a shared ancestor. The same process caused morphological and lifestyle divergence in the lineages containing each of these lichens. A relative of *Peltigera*, *Porpidia flavocaerulescens*, differs in body shape. In contrast, the closest extant relatives of the *Cora* lineage are moss-associated *Arrhenia* mushrooms (here *A. chlorocyanea*).

↓ A. *Peltigera canina;*
B. *Cora paraciferrii;*
C. *Porpidia flavocaerulescens;*
D. *Arrhenia chlorocyanea.*

DIVERGENCE

Morphological divergence is often observed in species-rich groups, and may arise after an evolutionary innovation that opens inroads to move into new habitats. An example is seen in the striking variety of fruiting bodies in the family Graphidaceae, all going back to a common ancestor likely with a simple pore, as depicted in the upper middle picture.

→   A. *Myriotrema microporum;*
B. *Phaeographis lobata;* C. *Stegobolus subwrightii;* D. *Gintarasia lamellifera;*
E. *Stegobolus radians;* F. *Glyphis cicatricosa;* G. *Dyplolabia afzelii;*
H. *Platygramme caesiopruinosa;*
I. *Allographa chrysocarpa;* J. *Platythecium grammitis;* K. *Sarcographa heteroclita.*

SALTATIONAL SHAPE CHANGES

Natural selection, as one Darwin meme popularized, involves "very slow change we can believe in." Animal body plans are incredibly conservative, governed by a cell-fate switchboard that goes back to the earliest forks in evolution. However, in some lichens the shape has switched completely within short evolutionary timescales for the participating fungus—a phenomenon known to evolutionary biologists as saltational change.

Saltational changes in lichens likely occur because of interference with signaling with the symbiotic partner, as opposed to slow changes in how hyphae are being arranged relative to each other. And saltational changes have one thing in common: they drive systematists and taxonomists crazy.

↗ Were it not for similar fruiting bodies (and DNA evidence), the thread lichen *Coenogonium interpositum* (left) would not be placed close to the crust lichen *C. strigosum* (right).

→ The fungi of *Punctelia* lichens (left) (here *P. subrudecta*) and their relatives form foliose lichens. Except one lineage, the one closest to *Punctelia* itself. And it's not that it has a different body plan, for it has no actual lichen "body." The *Nesolechia* fungus (here *N. oxyspora*) is but a fraction of the size of *Punctelia* lichens and forms reddish-brown, wart-like structures on other foliose lichens, which in reality are its fruiting bodies (right).

One fungus, two lichens

A symbiosis between multiple players, a lichen is perceived as a single organism, its players giving rise to forms not known from the individual components. It therefore seems logical that different-looking lichens are different organisms. The finding that they may involve one and the same fungus has vexed lichenologists, and vividly demonstrates that in symbiosis, the choir makes the music.

Two hundred years ago, the French lichenologist Dominique François Delise described a green algal lichen from the Canary Islands. It reproduced sexually with apothecia, and he named it *Sticta canariensis*. From the same region, he also described a cyanobacterial lichen reproducing asexually with isidia, *Sticta dufourii*. Few doubted the status of these as two distinct species, until a possible connection was postulated by the British lichenologist Peter James (1930–2014) in the mid-1970s. And then, in 1991, Daniele Armaleo and Philippe Clerc at Duke University used DNA sequencing techniques to prove that both lichens are formed by the same fungus but with different photobionts, a so-called photosymbiodeme, composed of different photomorphs.

Unknown to many, a similar connection had already been made at the beginning of the twentieth century by Alexander Zahlbruckner, between the green-algal lichen known as *Ricasolia amplissima* and its cyanobacterial counterpart described as *Dendriscocaulon*, a tiny shrub-like morph also known from green-algal lichens in the genus *Sticta*. We still do not know the signaling mechanisms that lead to these strikingly different architectures, but the switches are clear evidence that, in the business of lichen self-assembly, the photobiont matters.

Not all photosymbiodemes form disparate-looking lichens. In some cases, the green algal and darker hued, often gray, cyanobacterial morphs agree in overall shape. Also, associations of the same fungus with green algae and cyanobacteria do not always lead to photosymbiodemes. More common and widespread, perhaps representing the archetypal form of such associations, is the presence of both an alga and a cyanobacterium in the same lichen.

← The photomorphs of the New Zealand lichen *Pseudocyphellaria rufovirescens* do not differ much in gross morphology, both forming large, leafy thalli. Green algal and gray cyanobacterial lobes (the latter previously named *Pseudocyphellaria murrayi*) often grow intermingled.

↑ In the paleotropic genus *Gibbosporina* (here *Gibbosporina nitida*), the tripartite lichens are composed of a green-algal primary thallus, with the cyanobacteria confined to small lobules or branched cephalodia.

→ Tripartite lichens, in which the same fungus associates with both a green alga and a cyanobacterium, typically segregate the cyanobacteria in cephalodia, which can be borne as external flecks, as in the Flaky Freckle Pelt (*Peltigera britannica*, photographed on Cortes Island, British Columbia), or be harbored inside the thallus, as in the Lung Lichen (*Lobaria pulmonaria* – inset above, showing a false-colored scanning electron micrograph of the dense cyanobacterial colony in a thallus cross-section).

←↓ The two lichens depicted are formed by the same fungus, *Sticta latifrons* (a New Zealand endemic), the large foliose one associated with a green alga and the tiny shrubby one with a cyanobacterium. The cyanobacterial form looks so different it was originally classified in the genus *Dendriscocaulon*.

In these so-called tripartite lichens, the majority of the thallus consists of the fungus and an alga, while cyanobacteria are localized in discrete cephalodia, which may appear as warts, fleck-like or bulbous outgrowths (page 201), or bull's-eye rosettes (page 149) on the thallus surface. They can even be located inside the thallus, as in the Lung Lichen (inset on opposite page).

Some photosymbiodemes involve two different green algae instead of the "classic" green alga–cyanobacterium pairing. The sterile lichen *Buellia violaceofusca*, with a *Trebouxia* alga, has been shown to involve the same fungus as the fertile *Lecanographa amylacea*, in which the fungus consorts with a *Trentepohlia* alga. In all cases of photosymbiodemes, whatever their composition, the ability to consort with multiple photobionts enables the lichen fungus to occupy a broader range of niches and disperse over a larger area.

The architecture of co-dispersal

About a quarter of all lichens reproduce with particles that, upon magnification, reveal photobiont cells wrapped in a fungal swaddle. As we have seen, these support the lichen life cycle by co-dispersing the symbionts simultaneously. But they are also an architectural component of the thallus, providing openings into its interior or increasing its surface area.

Each of these particles is a just-add-water, ready-to-go lichen starter. They come in a variety of forms, but two basic motifs stand out. Tiny balls of fungal hyphae wrapped around clusters of photobiont cells, with a "soft" appearance, are called soredia. Cylindrical outgrowths with a hard "shell" (the thallus cortex) are called isidia.

Soredia tend to be formed in specific structures on the thallus, called soralia, which emerge from the inner photobiont layer and medulla and hence contrast with the remaining thallus. Dust or leprose lichens are entirely dissolved into soredium-like granules. Isidia often cover extensive portions of the surface of the lichen, taking more or less the color of the thallus, and while still affixed to the thallus they considerably increase its surface area.

Co-dispersal does not end with the standard, "over-the-counter" soredia or isidia. Nor do isidia or soredia come in only one shape or size. In some crustose lichens, isidia may become so large and so richly branched that the lichen assumes a minutely shrubby shape. Other lichens do not bother with isidia or soredia, but their thallus becomes so brittle that it fragments upon contact, with the resulting crumbs dispersed by mammals. Isidia may become flattened or form miniature thallus lobes, even including small rhizines, detaching from the thallus as miniature "start-up" lichens. Some lichens form isidia-like structures in the shape of disks, called schizidia, occasionally so numerous that the thallus surface appears to dissolve into rough scabs.

→ The near-cosmopolitan *Pseudocyphellaria citrina* (A) is known from all continents except Antarctica; it produces bright yellow soralia strongly contrasting with the brownish thallus. The tropical crust lichen *Acanthotrema brasilianum* (B) features rather robust isidia emerging from the thallus. In the cyanobacterial lichen *Coccocarpia prolificans* (C), the flattened isidia resemble minute scales.

↑ Soredia from the Hooded Rosette Lichen (*Physcia adscendens*). Each ball consists of "balloon-dog"-style fungal hyphae wrapping around a bundle of algal cells, which are in this case not visible to the outside (false-colored scanning electron micrograph).

↗ Isidia on the surface of the Lustrous Camouflage Lichen (*Melanohalea exasperatula*). The isidia are continuous with the surrounding cortex but readily break off and serve as all-in-one dispersal propagules; note the pits exposing algal cells within (false-colored scanning electron micrograph).

THE SPECIES-PAIR QUESTION

In early lichen research, taxonomists interpreted
differences in reproductive mode among otherwise
identical lichens as signaling the existence of different
species. A classic example of this is the recognition
of two near-identical species, one of which possesses
fruiting bodies such as apothecia, but no soredia, and
another that possesses soredia, but rarely fruiting bodies.
Such doubles have been termed "species pairs," and
they have been hypothesized to arise through the loss
of sexuality and emergence of asexual reproduction
in one fungal lineage. The phenomenon raises many
questions. Are species pairs really the result of two
lineages going down different evolutionary paths?
Does the loss of sexuality carry negative consequences?
Can an asexual species revert to sexuality?

As is so often the case, DNA sequencing of the
lichen fungus has shed light on these questions, and it
seems that there might not be a simple answer. In some
cases, sexually reproducing and asexually co-dispersing
entities have been confirmed as separate species,
sometimes closely and sometimes distantly related.
In others, a single fungal species turns out to be able to
do both, suggesting that the lichen develops one or the
other reproductive mode as needed. In many instances,
however, more than two species are involved, and
genetic species boundaries do not correlate with any
one reproductive strategy. An example is seen in
highlighter lichens of the genus *Letharia* (see page 274).
Starting out historically with the recognition of two
species, one with apothecia and the other with isidia,
taxonomists now recognize six, with either apothecia
or isidia, or with both.

→   The widespread lichens
Candelaria fibrosa (right, with
apothecia) and *C. concolor* (far right,
with soredia) form a well-known species
pair, sharing the same morphology but
differing in reproductive strategy. DNA
sequencing has revealed that more
than two fungal species exist in lichens
with this morphology, reproducing
either sexually or vegetatively or both.

Pored Net-coral Lichen

Latticework in nature

SCIENTIFIC NAME	*Pulchrocladia retipora* (Labill.) S. Stenroos, Pino-Bodas & Ahti
PHYLUM, FAMILY	Ascomycota, Cladoniaceae
GROWTH FORM	Mid-sized, fruticose, with densely pored branches
SPECIES IN GENUS	3
HABITAT	Coastal and alpine heathlands
NOTABLE FEATURES	Species of this genus are prime examples of latticework in nature

In construction, latticework is widely used for transmission towers, bridges, and greenhouses, but it also features in famous structures such as the Eiffel Tower. In nature, it is found at a microscopic level in diatoms—and at a larger scale in the net-coral lichens.

If there are lichens that epitomize the term "lichen architecture," it is the net-coral lichens of the genera *Pulchrocladia* and *Rexiella*. These lichens are a prime example of the use of latticework in nature to achieve stability with limited investment of biomass.

Net-coral lichens were for a long time included in the genus *Cladia*. DNA sequence data have shown, however, that *Cladia* can be divided into three evolutionary lineages, two of these showing a peculiar lattice structure and one with scattered, large pores along the branches. The latter are the true *Cladia* species, and so the other two needed new names: *Pulchrocladia*, found only in Australasia, and *Rexiella*, occurring also in the Andes of South America.

Common names are often ambiguous. The name "coral lichen" has been applied to several groups of fruticose, whitish lichens, due to their resemblance to branching corals, most commonly for species of *Pulchrocladia* and *Sphaerophorus* (page 236). But curiously, the first lichen to be dubbed a "coral lichen" is one described as *Lichen corallinus* by Linnaeus, now known as *Lepra corallina*—a crust lichen not related to either *Pulchrocladia* or *Sphaerophorus*.

→ *Pulchrocladia retipora* is considered by many to be among the prettiest lichens. It is found across Australasia, this particular specimen originating from southeastern Australia.

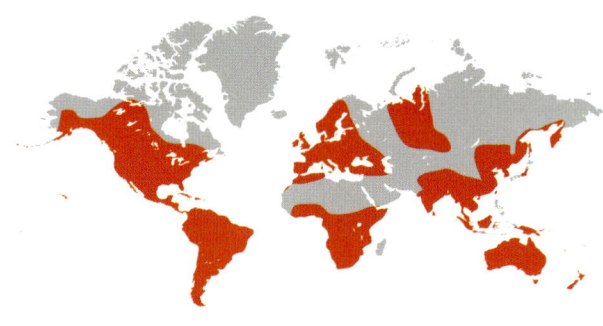

CHRYSOTHRIX CANDELARIS

Gold Dust Lichen

Explosions of fluorescent yellow dust

SCIENTIFIC NAME	*Chrysothrix candelaris* (L.) J. R. Laundon
PHYLUM, FAMILY	Ascomycota, Chrysothricaceae
GROWTH FORM	Thallus completely dissolved into soredia
SPECIES IN GENUS	17
HABITAT	Humid forests and sheltered rock outcrops
NOTABLE FEATURES	Covers the substrate in a bright yellow coating, and has been used for dyeing

In ecological terms, one of the least-appreciated lichen architectures is among the most successful. Dust lichens, also called leprose lichens (from *lepra*, dust, the same root word as in "leprosy"), are lichens in which all complex architecture is forgone, and the lichen consists only of masses of soredia.

Soredia are microscopic bundles of fungus with a bit of alga in the middle, and they have all that is needed to start a new lichen someplace else. You can think of them as lichen convenience-store packs—just add water.

Reducing your thallus to fungal–algal dust particles is a great way to get around, and it cuts out the gamble of having to find a new symbiont partner if you arrive on a new tree as, say, a spore. But it probably has downsides, too. Fungi like the one in the Gold Dust Lichen that are obligately asexual—that is, they don't produce fruiting bodies or spores—probably accumulate harmful DNA mutations over time, as they lack the customary sex-based mechanisms to eliminate them. More research needs to be done on this lichen lifestyle and whether or not it constitutes a successful last hurrah in evolution that is ultimately a dead end.

→ *Chrysothrix candelaris* coating spruce trunks and branches in a forest in the East Kootenay region of British Columbia, Canada. The species often gives the impression that a person has run amok with yellow spray paint.

Black-rimmed Byssus

Soft-shelled competitor

SCIENTIFIC NAME	*Sagenidiopsis undulata* (Fée) Egea, Tehler, Torrente & Sipman
PHYLUM, FAMILY	Ascomycota, Roccellaceae
GROWTH FORM	A crust lichen with a felty thallus forming a white–black marginal line
SPECIES IN GENUS	5
HABITAT	Tropical rainforest
NOTABLE FEATURES	The finely woven thallus is hydrophobic, so water pearls off

Crust lichens come in many forms, some thin and barely visible as a film over the substrate, others thick and conspicuously cracked. The byssus lichens are a peculiar type of crust lichen, their thallus being composed of numerous loosely woven hyphae and so appearing "*byssoid*" (from the Greek *byssos*, referring to the fiber used to weave linen).

Owing to this structure, byssus lichens repel water, so that the thallus does not remain damp all the time under highly humid conditions, facilitating gas exchange. And they easily overgrow other lichens.

Various genera include byssus lichens: besides *Sagenidiopsis*, also *Chiodecton*, *Crypthonia*, *Dichosporidium*, and *Herpothallon*. All of them are relatively closely related members of the *Arthoniales*, a diverse group of ascomycetes including lichens and fungi with other lifestyles.

Byssus lichens are predominantly found in the tropics, where it is warm and permanently humid. *Sagenidiopsis undulata*, found in Central and South America, can cover large areas of tree bark, individual thalli growing side by side and forming intriguing patterns with their white–black borderlines. While this species reproduces sexually with apothecia, a close cousin, *Sagenidiopsis isidiata*, reproduces with fluffy vegetative propagules. The two represent a so-called species pair (page 140), two lichens with the same underlying morphology but different reproductive strategies.

→ Numerous thalli of *Sagenidiopsis undulata* growing on the bark of a tree in southeastern Brazil. Apparently, this area of the bark was simultaneously colonized by numerous spores from a single dispersal event, several of them successfully establishing new lichens and then growing to produce this intriguing pattern of a lichen colony.

Rosy Bull's–eye Lichen

Best of both photobionts

SCIENTIFIC NAME	*Placopsis rhodocarpa* (Nyl.) Nyl
PHYLUM, FAMILY	Ascomycota, Trapeliaceae
GROWTH FORM	Medium-sized, crustose, with marginally radiating lobes and cephalodia
SPECIES IN GENUS	60
HABITAT	Cool-temperate to boreal and subantarctic zones and tropical mountains
NOTABLE FEATURES	Conspicuous cephalodia contain nitrogen-fixing cyanobacteria

The formation of morphologically distinct symbioses of the same fungus with either green algae or cyanobacteria is one of the most striking phenomena in lichens. Sometimes, the two photobionts are incorporated into the same thallus, the green alga then located in the algal layer and the cyanobacterium restricted to specific areas of the thallus, so-called cephalodia.

Cephalodia may be formed internally, invisible to the naked eye, but in some lichens they produce conspicuous external structures. In the genus *Placopsis*, the neat, small, radially lobulate cephalodia in the center of the thallus almost look like separate organisms invading another lichen. Their peculiar appearance gave *Placopsis* the name bull's-eye lichens.

Lichens that include two different photobionts, a green alga and a cyanobacterium, are also called tripartite lichens. Both photobionts are capable of photosynthesis, but the

cyanobacterium can do much more. Bacteria and archaea are the only organisms that can fix atmospheric nitrogen (N_2), converting it into ammonia (NH_3) and its derivates, which can eventually be used by other organisms—plants, animals, and fungi—to build amino acids, proteins, or nucleic acids (DNA). Nitrogen fixation is therefore essential to life.

Nutrient-poor habitats often lack sufficient nitrogen in absorbable form, and so the organisms living there are dependent on nitrogen-fixing bacteria or archaea. A cyanobacterial photobiont comes in quite handy in such situations, enabling the lichen to conquer habitats otherwise difficult to colonize. But why not just have a cyanobacterium as the photobiont? The presence of two photobionts, a green alga and a cyanobacterium, gives the lichen an additional advantage, as these photobionts deliver different goods to the fungus and operate under different microclimatic conditions, rendering the lichen more flexible under varying environmental situations.

Fixation of atmospheric nitrogen (N_2)

Fixation occurs in specific cells, heterocysts. In these, light is transformed through photosystem 1 into the cell's energy currency, ATP, which together with NADH and an enzyme, nitrogenase (only found in bacteria), converts N_2 into ammonium (NH_4^+). Ammonium is then used to build amino acids and other nitrogen-containing compounds (N-compounds).

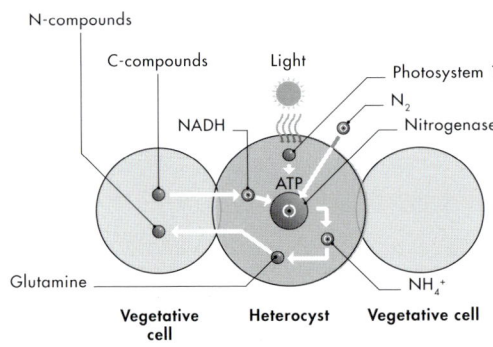

→ Thalli of *Placopsis rhodocarpa* growing on a rock outcrop in the Colombian paramo. The dark, radiating spots mark the cephalodia containing the cyanobacterium, whereas the rest of the thallus associates with a green alga. The pink disks are the apothecia, and the light green structures produce vegetative propagules.

Věžda's Leaf Dot

Half-moons in pristine rainforests

SCIENTIFIC NAME	*Badimia vezdana* Lücking, Farkas, & V. Wirth
PHYLUM, FAMILY	Ascomycota, Ramalinaceae
GROWTH FORM	Small crustose lichen with disk-shaped apothecia
SPECIES IN GENUS	19
HABITAT	Neotropical rainforest
NOTABLE FEATURES	Produces peculiar, half-moon-shaped fruiting bodies

Because leaves have a short life span and even in tropical rainforests rarely last for more than three years, lichens colonizing them need to complete their life cycles quickly. They do this by rapid and effective dispersal through unique reproductive structures.

One such structure, the often half-moon-shaped campylidia, produces fungal conidia that are dispersed through a rain-splash mechanism.

Campylidia evolved many times independently in unrelated ascolichen fungi. One of these, *Badimia*, is among the most spectacular elements of tropical phyllosphere communities. Its large thalli bear conspicuous, brightly colored apothecia and campylidia. The conidia produced by the campylidia consist of a filiform, curved main part in which each cell bears a small lateral appendage, adapted to float on water, which guarantees effective rain-splash dispersal and fast adherence to a new leaf surface once the water film dries off.

The almost 20 known species of *Badimia* are found only in pristine or well-preserved rainforest habitats, and serve as excellent indicators of environmental quality—that is, the preservation status of tropical rainforest areas.

Business on the side

Asexual reproduction in *Badimia vezdana* is achieved by means of peculiar spores produced in campylidia, halfmoon-shaped structures somewhat larger than the yellow apothecia. The spores are called conidia and are only about one micrometer thick; their small appendages help them to stick to the leaf surface after dispersal by rain water.

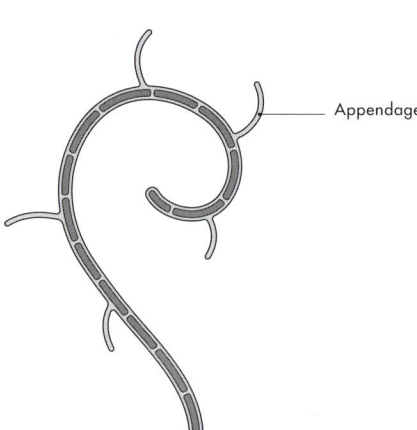

Appendage

→ *Badimia vezdana* covering parts of a leaf in Costa Rica. The round structures are the apothecia producing sexual ascospores, while the half-moon-shaped organs are the campylidia, which produce asexual conidia.

EVOLUTION
AND
TAXONOMY

Needles in a haystack: the fossil record of lichens

The study of ancient life and the reconstruction of extinct biota is one of the most fascinating branches of biology. Every year, hundreds of thousands of people line up to see dinosaur skeletons in museums. Long before dinosaurs, lichens may have played an important role in the evolution of early terrestrial ecosystems. Unfortunately, they and other fungi are not well represented in the fossil record, lacking the chemical attributes that facilitate fossilization.

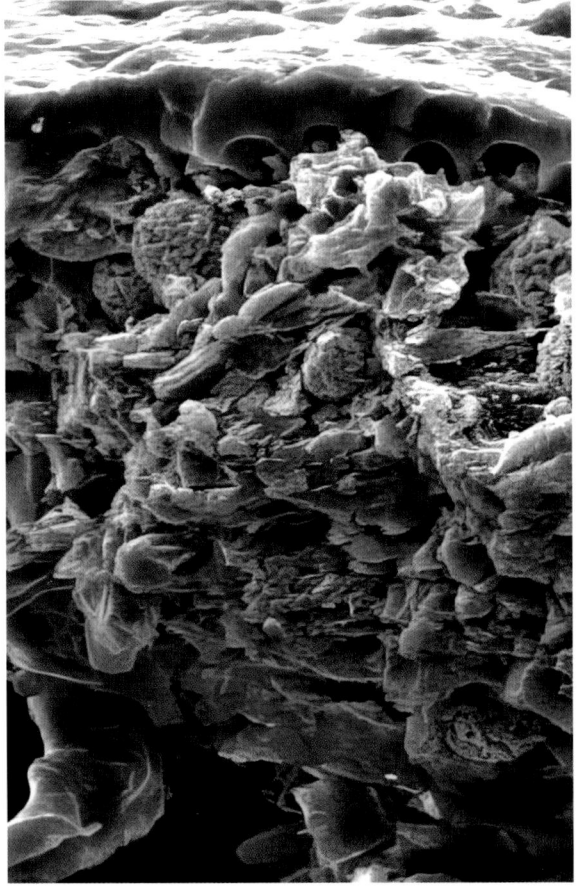

Billions of years of evolution are recorded in sediments in the form of fossils. Their first discovery upended our view of past life on Earth, and fossils remain crucial for building a timeline of how life evolved. Not least, they give us a glimpse into what extinct creatures looked like—their size, proportions, and features recorded in stone. Much of what we have learned is thanks to the deposition of hardened tissues such as bone and cartilage and the imprints left by plant leaves in clay and mud. By contrast, lichens (and other fungi) were preserved poorly in ancient rocks. While tens of thousands of fossils exist for animals and plants, fewer than 200 reliably dated and unambiguously assigned fossil lichens are currently known.

← The Swiss lichenologist Rosmarie Honegger described *Chlorolichenomycites salopensis*, the oldest known fossil accepted to represent a lichen.

↗→ *Daohugouthallus ciliiferus*, from the Jurassic period, is the oldest known lichen macrofossil. An artist's impression depicts it as an epiphytic lichen, with matching camouflage of the Mesozoic lacewing genus *Lichenipolystoechotes*.

HOW TO RECOGNIZE A FOSSIL LICHEN

Lichens do not have a well-defined and consistent body plan, and their fossils are not easy to recognize. There are only two ways a fossil can be reliably identified as lichen: either it resembles an extant lichen so closely that its identity is beyond doubt, or its internal anatomy exhibits the structure expected from a lichen, namely the intimate association of elements identifiable as fungal hyphae and algal or cyanobacterial cells. But assessment of the anatomy is challenging when dealing with compression, adpression, or impression fossils, in which the fossil represents the imprint of an organism but not the organism itself.

Many early fossils, going back over two billion years, have been identified as putative lichens. All putative lichen fossils up to the Precambrian, older than 539 million years, have been rejected or are classified as highly doubtful, including the famous Ediacaran "lichens" discussed on page 158. The oldest fossils generally accepted as lichens are from the Early Devonian (from a stage called the Lochkovian),

↖↗ The Ediacaran macrofossil *Dickinsonia costata*, depicted as fossil (Specimen P40135, South Australia Museum) and in an artistic rendering by paleoartist Nobu Tamura, is one of the most commonly cited putative lichens of the Ediacaran period. Its body plan and other features, however, support its interpretation as an extinct animal.

419–411 million years ago, one with green algae (*Chlorolichenomycites salopensis*) and one with cyanobacteria (*Cyanolichenomycites devonicus*).

After these, there is a gap of more than 200 million years, until the intriguing Jurassic macrolichen fossil *Daohugouthallus ciliiferus* (168–152 million years ago) and *Honeggeriella complexa* from the Early Cretaceous (about 133 million years ago). The latter pays tribute to Rosmarie Honegger, one of the world's foremost experts on lichen fossils. The majority of accepted lichen fossils are much younger, largely originating from Oligocene and Eocene amber between 20 and 40 million years old.

EARLY LICHEN HOPEFULS

An intriguing set of fossils that have been interpreted as lichens are the Ediacaran macrofossils from the Precambrian (635–539 million years ago). Ediacaran fossils encompass a diverse range of morphologies and are often considered early-diverging animals. The idea that some could represent lichens was put forward by Gregory Retallack, an expert on fossil soils. His argument was that the Ediacarans had high compaction resistance, which is inconsistent with their interpretation as soft-bodied animals. Lichens offered a convincing explanation, given the presence of chitin in their cell walls.

However, not only are there tougher substances facilitating fossilization, such as sporopollenin in plants, but chitin is also the main component of the exoskeleton of arthropods (and in fact only a small component of lichen fungal cell walls). If lichens (and other fungi) were indeed that tough, why is their fossil record so sparse? Furthermore, the Ediacarans typically have bilaterally symmetric body plans—unlike lichens—and lack an internal structure that could be interpreted as symbiotic interplay between fungal hyphae and algae or cyanobacteria.

The final verdict on the Ediacarans may not yet be in. We can rule out the charismatic macrofossils, but this does not preclude that lichen-like organisms might have lived in this time period. Diverse fungal fossils are known from the Ediacaran, and it was Retallack himself who described a lichen-like anatomy for the fossil *Petalostroma kuibis*. Some more recently discovered fossils do match the morphology of a radiating lichen thallus. *Nilpenia rossi*, for instance, comes remarkably close to the southern African endemic genus *Stellarangia*. But looking lichen-like may not mean much in the fossil world.

→ The Ediacaran macrofossil *Nilpenia rossi* bears a remarkable superficial resemblance to an extant lichen, *Stellarangia namibiensis*.

Placement of Ediacaran macrofossils

A cladogram, based on fundamental characters such as body-plan symmetry, shows the most likely placement of Ediacaran macrofossils (highlighted in boxes) within animals, plants, and fungi. Ediacaran fossils commonly cited as putative lichens invariably cluster with green, brown, or red macroalgae, or invertebrate animals such as jellyfish, flatworms, crustaceans, or sea urchins, but not with fungi (including lichens).

JURASSIC PARK RELOADED

Most fossils can be classified into compression, adpression, or impression fossils that reflect external morphology, permineralized fossils that allow analysis of internal anatomy, and amber fossils. The remarkable preservation of amber fossils, especially insects, has led to the idea that ancient DNA can be recovered, and the organisms perhaps recreated. Scientists have so far not been successful in recovering DNA from amber that is millions of years old, but the limits are continually being pushed back. To date, the oldest DNA to have been recovered hails from a two-million-year-old Pleistocene boreal ecosystem in Greenland, documenting ancient mastodons and reindeer

→　The best-preserved lichen fossils are found in Bitterfeld and Dominican amber from the Oligocene (20–30 million years old) and Baltic amber from the Eocene (34–40 million years old), such as this exquisite, undescribed *Cladia*.

↖ Species of *Umbilicaria* (here
U. calvescens, photographed in Peru) might
be the oldest known macrolichens, the
genus going back around 130 million
years. Whether ancient macrolichens
looked like the *Umbilicaria* lichens of today
is not known.

↗ Surviving non-avian dinosaurs,
macrolichens such as this gorgeous Lung
Lichen (*Lobaria pulmonaria*) from Norway
have evolved over the past 66 million years
as the dominant element of lichen
communities.

There and back again?
A fungal story

After biologists recognized their symbiotic nature, it did not take long to realize that lichens are not a genealogical lineage but represent a lifestyle found in different branches of the fungal (and algal) trees of life. Some branches diversified into thousands of species. Others remained small or suffered mass extinctions, their present-day representatives mere remnants of a greater ancient diversity. Still others evolved out of symbiosis, adopting other lifestyles. How often this back-and-forth happened is not fully resolved.

LICHENS EVOLVED MORE THAN ONCE

Perhaps the easiest question to answer is whether lichen symbionts evolved only once within the fungi. They did not. Although extant lichens are not known from outside the two largest fungal phyla, the Ascomycota (sac fungi) and the Basidiomycota (toadstools and relatives), their evolution within these two phyla was independent. Both groups contain early-diverging lineages that do almost anything except getting involved with lichens, such as the yeasts in the Ascomycota and the rust and smut fungi in the Basidiomycota.

The Basidiomycota contain only about 200 species of fungi involved in lichens. Yet their occurrence in three different, major groups within the Basidiomycota—as relatives of chanterelles, boletes, and agarics, respectively—makes it highly unlikely that all go back to a single origin of the lichen lifestyle. This means that the lichen lifestyle evolved at least three times in the Basidiomycota. In the Ascomycota, it is not quite that simple.

DID *PENICILLIUM* EVOLVE FROM LICHENS?

The fungi involved in the overwhelming majority of lichens, 99 percent, are found in the Ascomycota. This is the largest phylum within the true fungi, with nearly 100,000 species known, and 20,000 of these participate in lichens. The enormous diversity within this phylum is organized in several classes, a dozen of them characterized by forming conspicuous spore-bearing structures. The three best known are the Pezizomycetes, containing cup fungi, morels, and truffles; the Eurotiomycetes, containing the ubiquitous molds in genera such as *Aspergillus* and *Penicillium*; and the Lecanoromycetes, containing more than three-quarters of all fungi involved in lichens.

→   Multiple independent lichenization in basidiomycete fungi: (A) *Multiclavula mucida* is a relative of the chanterelles; (B) *Lepidostroma calocerum* is sandwiched between crust fungi and boletes; and (C) *Cora benitoana*, (D) *Lichenomphalia lobata*, and (E) *Dictyonema sericeum* are close relatives of agaric mushrooms.

The latter two classes are closely related, and both have a diverse biochemical arsenal at their disposal. Entertaining the thought that molds evolved from lichen symbionts is therefore not far-fetched. Indeed, a study published in 2001 first proposed the idea that Eurotiomycetes evolved from within Lecanoromycetes, and *Penicillium* at some point had a lichen ancestor.

New studies suggest that Eurotiomycetes and Lecanoromycetes have a close relative, the Lichinomycetes, that includes both lichen fungi and a range of enigmatic non-lichen fungi such as earth tongues (*Geoglossum*). These three classes together have been called the LLE group. The LLE group is related to another set of two classes, Arthoniomycetes and Dothideomycetes, the first consisting mainly, the second partially, of lichen fungi. Together, these five classes contain at least 11 lineages largely composed of lichen fungi.

It is not clear whether these lichen formers evolved independently, or whether some or all of them are the result of a single lichenization event potentially as far back as the Carboniferous, over 300 million years ago, followed by losses of the lichen symbiosis leading to important groups of non-lichenized fungi. That would be a mind-boggling scenario. Even if only the Eurotiomycetes evolved from lichen ancestors, it would mean that lichens gave rise to fungi as discomforting as the cause of athlete's foot (*Trichophyton rubrum*) and as diverse and useful as molds in the genera *Aspergillus* and *Penicillium*, including the most important antibiotic of the twentieth century, penicillin.

OTHER FUNGI WITH A LICHEN BACKGROUND

That non-lichenized fungi might have evolved from lichens is not just a theoretical scenario. One group within the Lecanoromycetes has become notorious for challenging long-held views on how lichen fungi relate to their non-lichenized counterparts. This group, a collection of related fungal families referred to in the taxonomic hierarchy as a subclass, is called the Ostropomycetidae, or ostropos for short. It contains over 4,000 species of lichen symbionts, as well as diverse saprotrophs, lichen parasites, plant parasites, and even some species that appear to be both lichen symbiont and saprotroph at the same time. Besides this impressive diversity of lifestyles, the ostropos defy established paradigms of morphology-based fungal systematics, as their members produce an unparalleled diversity of fruiting bodies and are not picky when it comes to lichen photobionts either.

A significant feature of ostropos is that, in contrast to the examples discussed above, little doubt exists that non-lichen fungi arose from lichen fungal ancestors. And the non-lichen fungi in this group often exhibit traits so divergent from those of their lichen-inhabiting relatives that they would never have been known to be related were it not for DNA sequence data. If saprotrophic leaf-surface fungi, such as in the genus *Rubikia*, evolved from lichens, then it might not be too far-fetched to think that *Penicillium* did as well.

→ Many fungi including lichen symbionts were once classified in distant groups, owing to their marked dissimilarities, but DNA sequence data have shown them to be closely related. Examples include *Graphis myrtacea* (A), quite unexpectedly the leaf-dwelling fungus *Rubikia evansii* (B) with its peculiar asexual spores (C), and *Gyalecta ulmi* (D).

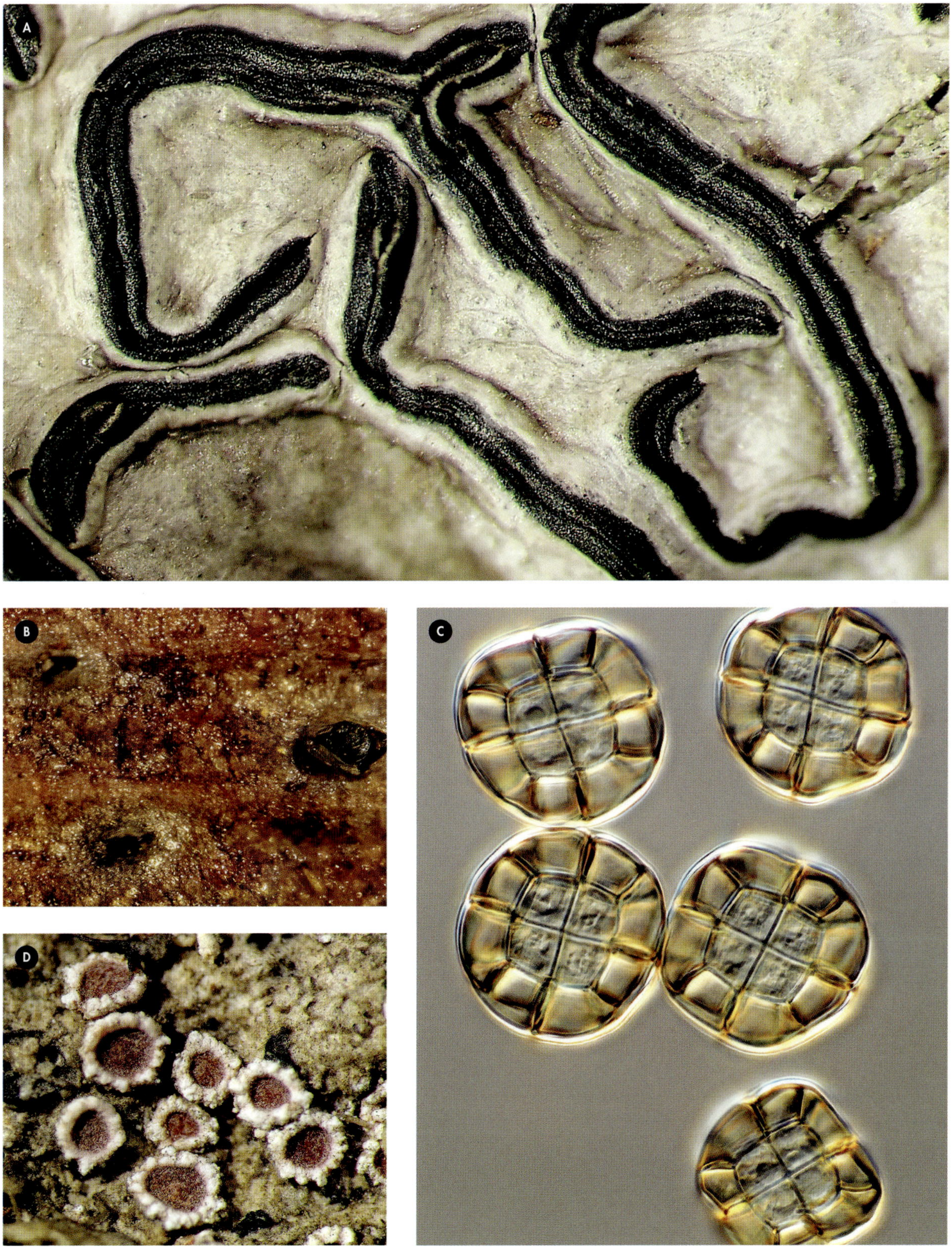

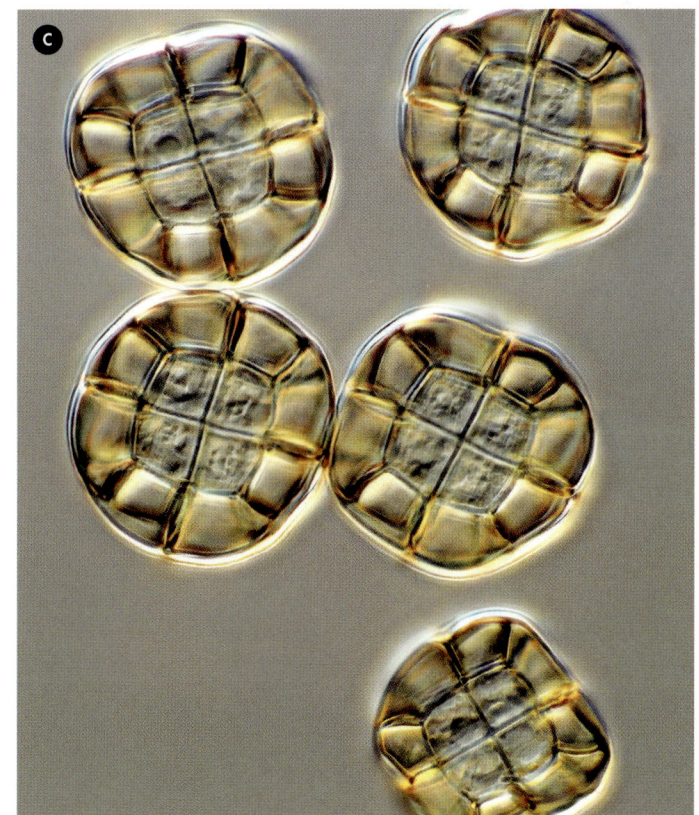

LICHEN FUNGI AND THEIR NON-LICHEN KIN

The closest relatives of lichen fungi are often not in lichens at all. The LLE lineage (Lichinomycetes, Lecanoromycetes, Eurotiomycetes) includes eighty percent of all lichen fungi, but it also contains fungi as diverse as earth tongues and Eurotiales molds, the latter of which play a major beneficial role in biotechnology. The orders Eurotiales and Chaetothyriales also include many harmful human pathogens.

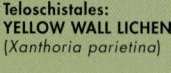
Teloschistales:
YELLOW WALL LICHEN
(*Xanthoria parietina*)

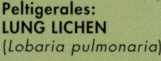
Lecanorales:
CHEWING GUM LICHEN
(*Protoparmeliopsis muralis*)

Peltigerales:
LUNG LICHEN
(*Lobaria pulmonaria*)

Rhizocarpales:
MAP LICHEN
(*Rhizocarpon geographicum*)

• GREEN SLICES ARE LICHEN FUNGI

LECANOROMYCETES

Graphidales:
FLESH SCRIPT LICHEN
(*Sarcographa labyrinthica*)

Ostropales:
WHITE-LOBED CUP FUNGUS
(*Stictis radiata*)

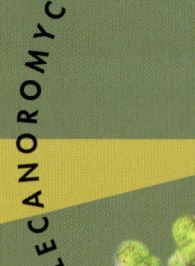

Trebouxia, a common lichen photobiont

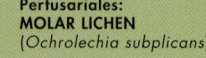
Pertusariales:
MOLAR LICHEN
(*Ochrolechia subplicans*)

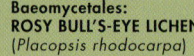
Baeomycetales:
ROSY BULL'S-EYE LICHEN
(*Placopsis rhodocarpa*)

Umbilicariales:
Bald Rock Tripe
(*Umbilicaria calvescens*)

Acarosporales:
BROWN COBBLESTONE LICHEN
(*Acarospora fuscata*)

Candelariales:
CANDLEFLAME LICHEN
(Candelaria concolor)

Geoglossales:
HAIRY EARTH TONGUE
(Trichoglossum hirsutum)

Symbiotaphrinales:
SMALL-FRUITED BEETLE GUT FUNGUS
(Symbiotaphrina microtheca)

Coniocybales:
Cottonwood Glasswhiskers Lichen
(Sclerophora amabilis)

• BROWN SLICES ARE NON-LICHEN FUNGI

LICHINOMYCETES

Lichinales:
TONGUE-BEARING SLIME CRUST
(Lempholemma lingulatum)

EUROTIOMYCETES

Verrucariales:
ELF-EAR LICHEN
(Normandina pulchella)

Chaetothyriales:
SAXOPHONE LUNG
(Exophiala phaeomuriformis)

Eurotiales:
PENICILLIN MOLD
(Penicillium chrysogenum)

Onygenales:
ATHLETE'S FOOT
(Trichophyton rubrum)

Pyrenulales:
CLUSTERED POX LICHEN
(Pyrenula anomala)

Bringing order to diversity

Before the recognition that lichens consist of both fungi and photobionts, naming them was a straightforward business. All the traits of the lichen, taken together, characterized that lichen, and it was given a scientific name. The name of a lichen species. The discovery of symbiosis created a conundrum—if the lichen was actually two or more organisms, what did the name apply to?

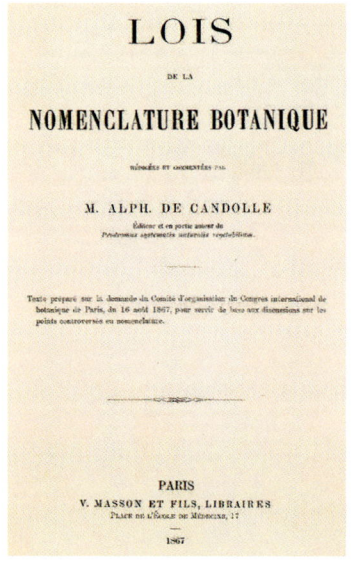

LOIS
DE LA
NOMENCLATURE BOTANIQUE

RÉDIGÉES ET COMMENTÉES PAR

M. ALPH. DE CANDOLLE
*Éditeur et en partie auteur du
Prodromus systematis naturalis regni vegetabilis.*

Texte préparé sur la demande du Comité d'organisation du Congrès international de botanique de Paris, du 16 août 1867, pour servir de base aux discussions sur les points controversés en nomenclature.

PARIS
V. MASSON ET FILS, LIBRAIRES
PLACE DE L'ÉCOLE DE MÉDECINE, 17
1867

↑ The first Code governing the nomenclature of plants, algae, and fungi was published in 1867, in the very same year Schwendener presented his theory about the lichen symbiosis.

→ These two lichens look quite different, and their fungi have different names (*Cladonia bellidiflora* versus *C. polydactyla*), but DNA evidence suggests they involve one and the same species of fungus.

THE NAME GAME

The naming of plants, algae, and fungi has been governed by a rule book, the "Code," since 1867. One of the rules built into the Code in 1906, further adjusted in 1935, was the seemingly self-evident provision that a name could only apply to a single organism. The discovery of the lichen symbiosis, in precisely the same year, along with the realization that fungi are not plants, complicated things.

Lichens had been named as standalone organisms since before Linnaeus, and after 1867, it was no longer clear which component the name referred to. Without explicitly agreeing on it, most workers applied the scientific name given to lichens to their fungal component, applying separate names to the photobionts, an approach also adopted in a large work on lichens by Alexander Zahlbruckner in 1907, only one year after the new naming rule was taken up.

Still, trouble brewed. In the late 1940s, a mycologist working at the New York Botanical Garden pointed out that the strict application of the "no mixture" rule would require the rejection of all names given to lichens where the photobiont had been described as an intrinsic part of the organism—in particular, names established before Schwendener. As a solution that preserved those names without needing to go back and rename them from scratch, an addendum to the Code was crafted that grandfathered in the existing names of "lichen species." The article, entering the rule book in 1952, states: "For nomenclatural purposes names given to lichens shall be considered as applying to their fungal components." Each lichen "species" became a fungal species. Snap.

The upside to this solution is stability to the taxonomic naming system, which is inarguably desirable. But was it a fully informed decision? Had scientists at the time checked that each "lichen species" could really be equated with one fungal species? Or could one fungal species be involved in more than one lichen symbiosis? What if what we see as a lichen is not merely the realized body plan of a fungal species but instead the outcome of multi-player interactions? Such is the case with photosymbiodemes, which may look dramatically different, yet their fungi, and hence their scientific names, are the same. The name *Sticta latifrons* could refer either to the conspicuous, foliose green lichen or to the tiny, easily overlooked, minutely shrubby cyanolichen (page 134).

The advent of DNA-based studies revealed yet more cases where the same fungal species participate in different, distinct lichen symbioses. One example includes the so-called "phantom phenotypes," lichens involving the same species of fungus but in which the photobiont does *not* differ. Even more confounding are the lichens that appear to uniformly share a predictable suite of morphological traits—what some lichenologists might call a "good species"—yet in which different thalli are inhabited by different (though usually related) fungal species. These have been called "cryptic species," for the reason that the fungal species cannot be reliably predicted from the lichen's appearance alone. Unnervingly, many such species—as now defined by the naming rules— cannot be reliably identified without DNA sequencing. Much ink has been spilled over these observations, but the examples cited above share a common thread that the lichen symbiotic outcome does not neatly line up with the one lichen, one fungus paradigm.

→ The fungus of the Gnomes' Bones Lichen (*Thamnolia vermicularis*), which has a truly worldwide distribution, was believed to represent a single species, until studies by Romanian-Swedish lichenologist Ioana Brännström and her team unveiled the existence of two additional species, one in the far north (*Thamnolia tundrae*) and one in the European Alps (*Thamnolia taurica*). The lichens involving these three fungal species are virtually indistinguishable cryptic species.

↑ Alpha taxonomy involves the accurate recognition of species and their placement into the correct genera. In the taxonomy of lichen fungi, major advances have been marked by technological breakthroughs. The microscope allowed workers to recognize the importance of spores. Secondary chemistry developed as a tool to characterize genera and species, brought to perfection by the likes of Chicita and Bill Culberson, two of the foremost lichen chemists. DNA sequencing turned our understanding of genera and species upside down. The lichen now known as *Polycauliona candelaria* (based on DNA sequence data) was classified in no fewer than 15 genera, including as *Lichen candelarius* by Linnaeus, *Lecanora candelaria* by Acharius (based on morphology), *Callopisma candelaria* by Trevisan (based on spores), and *Xanthoria candelaria* by Theodor Magnus Fries (based on morphology and pigment chemistry), before finding its current home based on fungal DNA sequencing.

↑   The common and widespread *Parmelia saxatilis*, the Salted Shield Lichen, was long thought to be a readily recognizable species, growing mostly on rock in temperate regions of both hemispheres. DNA sequencing revealed at least four "hidden" species within a comparatively small part of its global range. In many other cases, species we thought we knew turned out to hide unrecognized species, sometimes at an order of magnitude. The current world record is held by the basidiolichen genus *Cora*, until three decades ago believed to include a single species and now known to encompass at least 250 (see page 188).

LICHENS HAVE NO NAMES

In an ironic twist to the now century-and-a-half study of lichen symbioses, lichenologists increasingly acknowledge that the outcomes of the symbioses, the lichens themselves, have no names. The scientific names formally apply to the fungus and the alga. Often the lichen lines up with the fungal species, but surprisingly often it does not, or only with much hand-waving. But old traditions die hard: lichen fungi continue to be described in scientific literature based on their morphology in the symbiotic state, and many lichenologists still commonly equate fungal species with lichen species.

In this book, we frequently refer to the scientific name of the fungus as being *involved in* a certain lichen, and refer to its symbiotic outcome—aka the lichen—as the "(fungal name) lichen" or use the lichen's English name. Language matters, and we are confident that a little added pedantry will pay off by placing fewer expectations up front on the role of the fungus in the emergence of lichen outcomes, a role about which we still know very little.

INCORPORATING LICHEN FUNGI INTO THE FUNGAL TREE OF LIFE

Prior to the advent of DNA sequencing, fungi including lichen fungi, were classified based on a combination of morphological and anatomical characters, such as the shape and internal structure of the fruiting bodies, including the hymenium and the spore-bearing asci or basidia, and the type of spores. Throughout history, these characters were perceived and weighted differently, so that different systematists arrived at different solutions. Starting in the second half of the twentieth century, classifications converged on more broadly accepted schemes.

And just when the scientific community thought that a good classification had been obtained, along came DNA sequencing and threw everything overboard. Those defending traditional classifications argued that there must be something wrong with DNA data, but it soon became clear that much of what we had assumed about the evolution of fungal morphologies, including those of lichens, was wrong.

To cut a three-decade-long story short, the modern classification of fungi, including those involved in lichens, looks very different from what systematists had in mind only three decades ago.

↓   Naming lichens is not straightforward. The Fringed Rosette Lichen incorporates at least three names: the common name Fringed Rosette Lichen, applied to the lichen as a whole, the scientific name *Physcia tenella*, referring only to its fungal component, and the scientific genus name *Trebouxia*, denoting the green algae found in this lichen.

Clustered Pox Lichen

A tropical discovery going back to Acharius

SCIENTIFIC NAME	*Pyrenula anomala* (Ach.) Vain.
PHYLUM, FAMILY	Ascomycota, Pyrenulaceae
GROWTH FORM	Crust lichen with black spots of aggregate, partly immersed perithecia
SPECIES IN GENUS	170
HABITAT	Tree bark in tropical rainforests
NOTABLE FEATURES	One of the commonest and most easily recognized species of the genus

While foliose macrolichens are a familiar sight in temperate regions, tropical rainforests have an abundance of crust lichens. Many of these belong to groups where only a few species made it outside the tropics. The Pyrenulaceae are among these, and *Pyrenula anomala* is one of that family's most common members.

Pyrenulaceae are crust lichens, reproducing by means of spores produced in closed perithecia. Members of this family have black perithecia, but this is not sufficient for definitive identification: microscopic spore details are required. Spores of *Pyrenula* are brown, whereas those of a superficially similar group of tropical crust lichens, Trypetheliaceae, are colorless.

Pyrenula anomala is one of the few tropical crust lichens first described by Erik Acharius. Acharius mostly worked with temperate lichens and focused on their morphology, as his microscope was not good enough to observe spore details. Not aware of its peculiar spores, he placed the species in the genus *Trypethelium*, but gave it the name *anomalum* as it looked different from other species of that genus. After its true home in the genus *Pyrenula* was established a century later, it became clear that it is not an "anomalous" *Pyrenula*, as the name might suggest.

→ *Pyrenula anomala* is a common sight on tree bark in tropical rainforests. Most records are from tropical America and Asia, and the species has only rarely been found in tropical Africa. Spores of *Pyrenula* are brown, and in some species, such as *Pyrenula subpraelucida*, become quite large and look spectacular.

Compartmented Fleshscript Lichen

Intricate patterns for spore dispersal

SCIENTIFIC NAME	*Sarcographa heteroclita* (Mont.) Zahlbr.
PHYLUM, FAMILY	Ascomycota, Graphidaceae
GROWTH FORM	Crustose with script-shaped, divided apothecia
SPECIES IN GENUS	41
HABITAT	Tropical rainforest
NOTABLE FEATURES	The grouped apothecia are divided into numerous minute chambers

Script lichens of the family Graphidaceae represent the second-largest group of lichen fungi and the largest in tropical regions, with well over 2,000 species known and probably another 1,500 awaiting discovery. The intricate shapes of their fruiting bodies capture the attention of the observer, especially when many different species grow in close proximity on tree bark. Why they display such a diversity of shapes in their reproductive structures is unknown.

Sarcographa is one of about 80 genera currently distinguished in the Graphidaceae. Prior to the advent of DNA sequencing, the classification of this family was much simpler: just eight genera were distinguished, depending on whether their apothecia were dispersed or grouped in raised thallus areas, whether their spores were colorless or brown, and whether those spores were compartmented ("septate") in one direction only (transverse) or in both (muriform).

This schematic classification resulted in $2 \times 2 \times 2 = 8$ possible combinations. Easy to learn, but far from the reality of evolutionary relationships. *Sarcographa* was characterized by grouped apothecia and brown, transversely septate spores. The genus has survived the DNA revolution but has been expanded to include some species with muriform spores.

→   Species of *Sarcographa* are recognized by their apothecia grouped in raised, white areas, contrasting with the green to brown thallus. *Sarcographa heteroclita* occurs in all tropical regions and can be recognized by its elongate groups of apothecia that are divided into minute chambers.

GRAPHIS SCRIPTA

Common Script Lichen

Taxonomic mysteries written in bark

SCIENTIFIC NAME	*Graphis scripta* (L.) Ach.
PHYLUM, FAMILY	Ascomycota, Graphidaceae
GROWTH FORM	Crust lichen with whitish thallus and script-like apothecia
SPECIES IN GENUS	200
HABITAT	Smooth-barked trees in temperate forests of the northern hemisphere
NOTABLE FEATURES	The only common script lichen in temperate regions

The Common Script Lichen, *Graphis scripta*, is a member of the illustrious club of lichens treated in Linnaeus's *Species Plantarum* from 1753, as *Lichen scriptus*. Its name stems from the peculiar shape of its spore-producing apothecia, which are elongate and often branched patterns that recall hieroglyphics or runes.

Script lichens (Graphidaceae) are an essentially tropical family, with hundreds of species occurring in tropical forests. Only a few have made it into temperate and boreal regions, and among these, *Graphis scripta* is by far the most common.

One reason might lie in its photobiont, a species of *Trentepohlia*, a genus that is abundant in lichens of tropical habitats. Perhaps *Graphis scripta* associates with a particular strain of *Trentepohlia* better suited to tolerate cooler conditions. A similar case is found in other tropical crust lichens, the Trypetheliaceae. A single species, *Viridothelium virens*, has made it far into North America, and it may be that it associates with the same photobiont as *Graphis scripta*.

Graphis scripta was long believed to constitute a single species, widespread in the northern hemisphere, but fungal DNA sequencing paints a daunting picture: *Graphis scripta* consists of many genealogical branches, some so distinct that they likely constitute different species. However, there is little correlation with morphological characters, challenging suggested alternative classifications. For now, taxonomists are treating it as a "species complex".

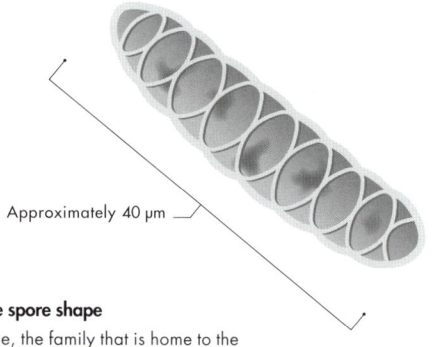

Approximately 40 μm

Unmistakable spore shape
Graphidaceae, the family that is home to the Common Script Lichen, includes more than 2,000 known species. Many have this peculiar type of sexual spores (ascospores), with numerous lens-shaped cells arranged in a straight row separated by thickened walls.

→ The Common Script Lichen, *Graphis scripta*, constitutes a "classic" lichen known to Linnaeus and workers before him. What appears to be a single species might actually be many.

Elephant Heart Lichen

Hundreds of species hidden in plain sight

SCIENTIFIC NAME	*Cora elephas* Lücking, B. Moncada, & L. Y. Vargas
PHYLUM, FAMILY	Basidiomycota, Hygrophoraceae
GROWTH FORM	Foliose with rounded lobes
SPECIES IN GENUS	About 90, but 450 estimated
HABITAT	Mostly in neotropical montane rainforests and paramos
NOTABLE FEATURES	The gray lobes are bordered by a thick white margin

DNA sequencing has revealed the phenomenon of hidden diversity: previously unrecognized species in what was believed to be a single, well-known species. When such "hidden" species are closely related, the connoisseur speaks of "cryptic speciation," not to be confused with being cryptic due to camouflage, but in a figurative sense—meaning something that is hidden in plain sight.

The current world record holder in the hidden-diversity sweepstakes is the genus *Cora* (heart lichens), with a single species that was accepted for a long time but is now known to represent hundreds of species.

When an international team of researchers looked at their DNA sequence data of *Cora* lichens, they first thought that there must be an error, as the data did not reflect what was believed to be a single species. Eventually it became clear that there was no error: what previously was thought to be a single *Cora* species turned out to be many different species. So far, about 90 have been enumerated and 160 more await formal description—and there seems to be no end in sight.

For years, *Cora* led a shadowy existence alongside its many ascolichen cousins, as one of just a few basidiolichens seemingly in existence. Studies by Manuela Dal Forno and Bibiana Moncada have now catapulted it into the top ten of the most speciose genera of lichen fungi, with an estimated 450 species.

→ *Cora elephas* is one of hundreds of previously unrecognized species of heart lichens. It was discovered in the paramo of Sumapaz in Colombia. Its name stems from its large size, robust appearance, and gray color, resembling an elephant.

LICHEN
ECOSYSTEMS

Conditions that favor lichen biomass

A significant proportion of the Earth's terrestrial surface—probably somewhere between 5 and 10 percent—is dominated by lichens. In practical terms, this means that in these areas the majority of cover on any given acre of ground would have greater cover of lichens than of vascular plants or mosses.

While 5–10 percent of the ground surface is impressive, focusing on the ground risks missing the bigger picture. In a much larger proportion of the Earth's vegetation, including its vast forests and even urban areas, lichens coat the vegetation in a third dimension and reach out into the air, achieving surface areas that in some cases rival those of the trees themselves.

LICHENSCAPES

The parts of the Earth's surface that are dominated by lichens tend to be environments humans would consider harsh. Lichen-dominated surfaces occur at all latitudes and altitudes. Though no one formula captures all of the environmental factors involved in setting up lichen domination, there are some recurring themes.

→ Lichenscape: Canada's Hudson Bay lowlands, here in Wapusk National Park, are dominated by mat-forming *Cladonia* lichens as far as the eye can see. On the map (inset) of North America and Greenland produced from the European Space Agency World Cover Project, the vast expanses of *Cladonia*-scapes are shown in pale yellow.

Lichens tend to be abundant in cold places with low evapotranspiration (evaporation plus transpiration by plants), such as tundra. High mountaintops, Arctic fellfields and Antarctic deserts are rich in both lichen species and lichen biomass. With annual mean temperatures frequently below freezing, they are inhospitable for much of life, but lichens can make a fine, if slow-paced, living here.

Lichens can also dominate where water can be extracted from the atmosphere in the form of fog or dew, but soils are too shallow or too dry for vascular plants. This favors the development of "lichenscapes" in unexpected places like coastal deserts. In the Namib Desert along the southwestern coast of Africa, whole plains are covered in shrub lichens with striking orange pigmentation, which likely shields the algae from intense sunlight. Here, sand surface temperatures can soar to 167 °F (75 °C), and lichens live off of morning fog generated from the cold waters of the nearby Atlantic Ocean. In short, lichens dominate in all of the places where their superpowers—their ability to extract what they need from the air, and their desiccation tolerance—give them an advantage over plants and mosses.

↖   Northern conifer forests are the most extensive terrestrial ecosystems, with a high diversity and abundance of lichens, often both on the ground and in the tree canopies. Whiteshell Provincial Park, Manitoba, Canada (top). Reindeer lichens (various species of *Cladonia*) cover extensive areas in boreal forests and Arctic tundra (bottom).

→   Tropical rainforests, such as in Henri Pittier National Park in Venezuela (A) or Madre de Dios in Amazonian Peru (B), support some of the highest small-scale diversity of lichens on Earth. In most plants and animals, related species are either ecologically differentiated or geographically separated. Lichen communities, in contrast, often harbor species of the same genus growing side by side. Here (C), six species of *Astrothelium* border each other on a small area of tree bark in Amazonian Peru.

→ Andean paramos (here the Páramo El Ángel in northern Ecuador) harbor rich lichen communities, but the diversity of lichens in these ecosystems is only appreciated upon closer examination.

← Leaves in a tropical rainforest—such as these in French Guiana—may be covered with diverse, colorful communities of tiny lichens hardly more than a few millimeters across.

↓ Leaf-dwelling lichens hold the record for the most diverse lichen communities at small scales, with up to 50 species recorded on a single leaf. This corresponds to 1 percent of the diversity of lichens found in entire countries, such as the United States or Brazil.

VEGETATION COATED IN LICHENS

Most of the vegetated surface of the Earth consists of forests and grasslands, where water supply and temperature allow trees and grasses to achieve their full potential. Lichens do not dominate here, but they are nonetheless major players. Indeed, it is here that lichens achieve some of their greatest diversity.

In addition to soil and rocks, the world's forests offer a mind-boggling diversity of organic substrates: bark and wood of tree trunks, branches, twigs, and, especially in the tropics, leaves. Further substrates are offered by their death and decay, including snags (standing dead trees), barkless branches, and logs. Substrates can be exposed, as in treetops, or shaded on the forest floor; they can be caressed by the breeze or sheltered in muggy calm air. Each plant species that supports lichens has its own bark and wood chemistry, the possible combinations generating a bewildering array of microsites.

Lichen biomass is not necessarily a predictor of species diversity, and some of the areas with the highest diversity might not even give the impression to the lay observer of lichen habitat at all. Such is the case with many tropical rainforests. Here, habitats that would elsewhere host lichens are smothered by bryophytes and epiphytic plants, but on the surfaces of evergreen leaves, lichens achieve astonishing levels of diversity, employing a combination of short life cycles and myriad dispersal adaptations to get around (page 53).

↑   The Namib Desert in southwestern Africa (Angola, Namibia, and South Africa) is known for its colorful lichen communities, mostly representing brightly pigmented members of the family Teloschistaceae, here *Dufourea flammea*.

↑   Antarctica harbors the most extensive cold deserts on Earth, and much of the macroscopic life there consists of lichens. Over 300 species have been recorded, mostly crust lichens on rock. Some even grow within the rock, where they are afforded protection from the elements. Among the more striking macrolichens is the iconic *Usnea aurantiacoatra*.

Water and nutrient cycling

Nitrogen cycle
Lichen-associated cyanobacteria (lower left) contribute a large amount to the fixation of atmospheric nitrogen, especially in habitats such as forests where other nitrogen fixers can be scarce.

It has long been recognized that lichens function as pioneers in primary succession, aiding in soil stabilization and facilitating the development of later plant communities. However, lichens contribute much more, forming important components of the water and nutrient cycles in terrestrial ecosystems.

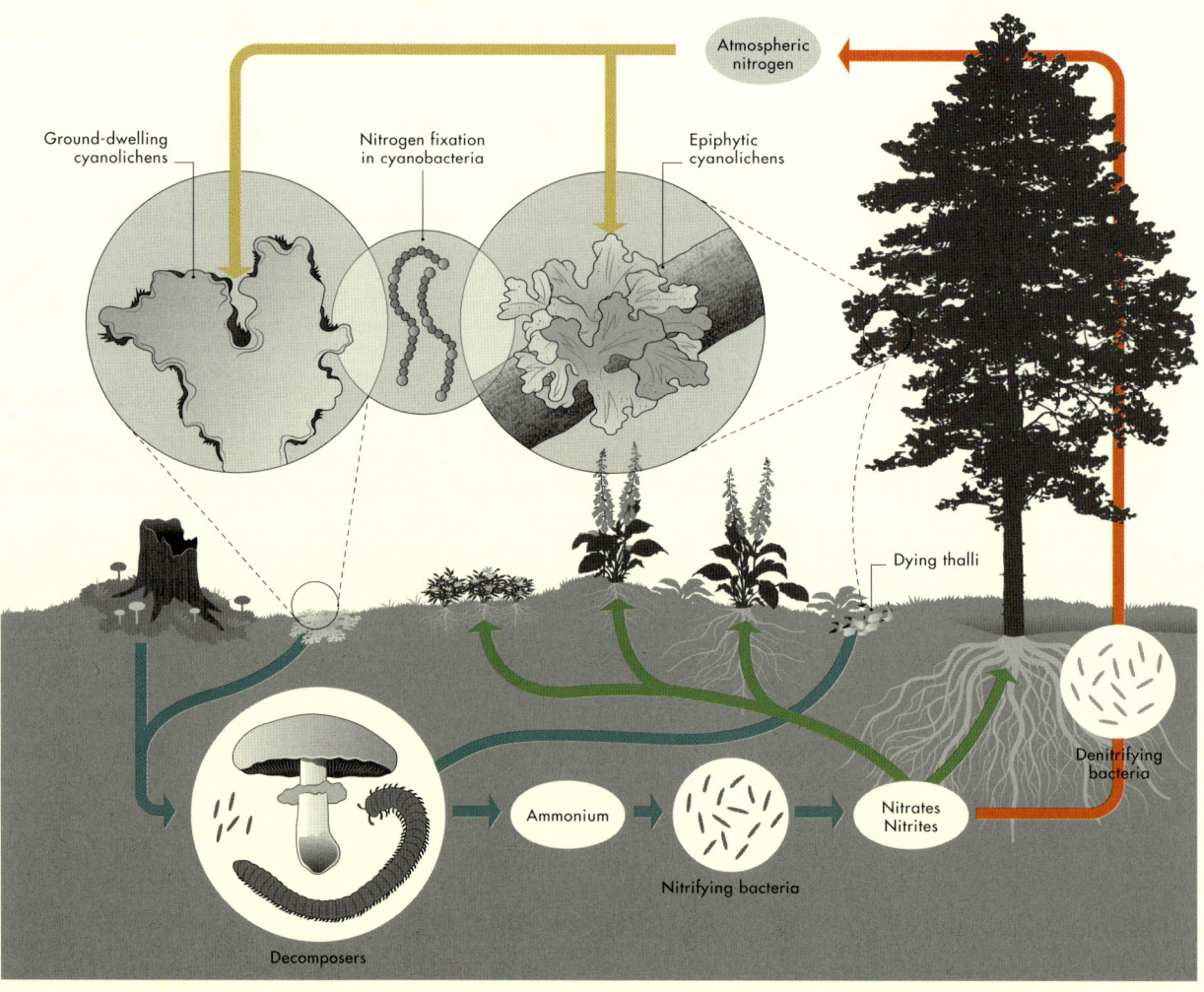

FIXATION

DENITRIFICATION

NITROGEN UPTAKE

MINERALIZATION

Atmospheric nitrogen

Ground-dwelling cyanolichens

Nitrogen fixation in cyanobacteria

Epiphytic cyanolichens

Dying thalli

Denitrifying bacteria

Decomposers

Ammonium

Nitrifying bacteria

Nitrates Nitrites

ABSORBING WATER LIKE A SPONGE

One difference between lichens and other fungi—and most plants—is that they can completely dry out, including at the cellular level, and then rehydrate and continue their business as if nothing happened. At the habitat scale, thousands of gallons of water are soaked up by lichens, holding water for slow release. This can be important in ecosystems as diverse as deserts, where soil-dwelling lichens mitigate soil erosion, and forests, where lichens help the ecosystem stay humid long after it has stopped raining, as has been demonstrated experimentally in a study in Costa Rica.

CONTRIBUTING TO THE NITROGEN CYCLE

While living organisms require all kinds of nutrients to survive, perhaps none is as important as nitrogen. It is a necessary building block for many essential molecules, including DNA and amino acids. Plants get their nitrogen from the soil in the form of ammonia (NH_3) or nitrate (NO_3). Animals and fungi obtain nitrogen through uptake of small organic molecules. While nitrogen is the most abundant element in our atmosphere, only bacteria are capable of converting inorganic atmospheric nitrogen into ammonium and nitrate. The nitrogen cycle thus depends on these usually invisible organisms.

Well, not entirely invisible. As we have seen, some bacteria are macroscopic and are capable of photosynthesis: the cyanobacteria. And here, lichens come into play, as a large proportion of terrestrial cyanobacteria are bound in lichen symbioses. It is not for nothing that nutrient-poor habitats exhibit an abundance of cyanolichens or green algal lichens with additional cyanobacteria, which contribute substantially to the nitrogen cycle.

↗   Jelly lichens (here, *Leptogium phyllocarpum*) are a prime example of the capacity of lichens to absorb water up to 500 percent of their dry weight and to keep it there for some time, acting as buffers to maintain atmospheric habitat humidity and prevent soil erosion.

→   Shrub lichens of the genus *Stereocaulon* (here *S. novogranatense* from Colombia) are a common sight in nutrient-poor habitats. The raisin-shaped cephalodia contain nitrogen-fixing cyanobacteria.

What's for dinner?

Lichens feature in the diets of hundreds of species around the world, from microscopic mites to the charismatic grazing megafauna of the Arctic. The role of lichens as keystone species in ensuring ecosystem function is only beginning to be understood, but it is increasingly clear that, without them, many other species would disappear.

Flying from Europe to the central United States, you may see the vast expanses of Canadian lichens on your in-flight tracker. Covering a thin layer of soil over rock, with only narrow ribbons of trees in between, these are *Cladonia*-dominated landscapes made up of reindeer lichens, termed "lichen barrens" by early European explorers, who could not see anything to monetize. Only 4–8 in (10–20 cm) tall, the reindeer lichens that live here plod along with lichen-like persistence, growing at their tips up to ¾ in (2 cm) a year. This might not seem like much, but growth is growth, and if they were not eaten or broken down, they would someday dome up to a tremendous thickness. But, like shrubs on the Serengeti, they are kept in check by grazers.

→ Reindeer lichens in the boreal northern hemisphere are the preferred winter food of Reindeer or Caribou.

MICROSCOPIC LICHEN FEEDERS

Most of the feeding action happens at microscopic scales, but its importance should not be underestimated. If you were transported into this lichen world, *Gulliver's Travels*-style visiting Brobdingnag, you would witness alarming feeding frenzies—large, ambling grazers and ambushing predators. In tropical rainforests, the surfaces of living leaves can be scenes of absolute destruction. The larvae of barklice (Psocidae) emerge after hatching on the leaf undersurface and feed on the lichens and algae growing on the upper surface, obliterating entire lichen communities in a few days. Elsewhere in this leafy world, nematodes may advance in carpets of slime, ingesting every lichen in their reach. Lichens on almost any substrate may exhibit zigzag "paths" over the thallus, left behind by the tiny oscillating heads of snails and slugs.

Often, lichenologists see the results but are unsure of the culprits. In some lichens, the starch-rich apothecia are hollowed out, spores and all, perpetrators unknown, long since having quit the scene. Cut a lichen open, and you might discover lichen-eating mites and those most charismatic of tiny invertebrates, the tardigrades (water bears). The lichen-visiting Gulliver may even get to watch them sucking out the contents of single algal cells.

→ A Yunnan Snub-Nosed Monkey eating what appears to be a Methuselah's Beard Lichen (*Usnea longissima*).

←↓ Feeding frenzy. Barklice are capable of eating away entire leaf-dwelling lichen communities.

SUSTAINERS OF MEGAFAUNA

The best-known lichen eaters are orders of magnitude larger. One of the charismatic hoofed mammals of the northern hemisphere, *Rangifer tarandus*, variously called the Caribou (in North America) or Reindeer (in Eurasia), is a lichen specialist *par excellence*. These animals are specially adapted to digest a lichen-majority diet, and their vision extends into the ultraviolet range, likely enabling them to see differences between the lichens that are invisible to us. Tundra and boreal-forest Caribou and Reindeer survive the winter by pawing through shallow snow to reach the lichens, and some achieve their greatest weight gain during this season.

Mountain Caribou in British Columbia, Canada, are uniquely adapted to standing atop 10 ft (3 m) snowpacks in winter to feed upon hair lichens of the genus *Bryoria*.

A smaller ungulate, the vampire-fanged Siberian Musk Deer (*Moschus moschiferus*), likewise depends almost exclusively on lichens for its winter survival. Even common ungulates such as the American White-tailed Deer (*Odocoileus virginianus*) will indulge in a lichen snack if offered a wind-fallen tree.

Having to tough out the winter as a herbivore in places with little else to eat might be just what has led some species to switch to a lichen diet. Like Canada's Mountain Caribou, the Yunnan Snub-nosed Monkey (*Rhinopithecus bieti*), endemic to southwestern China, gets through harsh mountain winters by eating hair lichens such as *Bryoria nepalensis* and *Usnea longissima*. Also like the Mountain Caribou, the monkeys are critically endangered, the result of poaching and habitat loss—the logging of the old forests required to support their hair-lichen diet. In the case of the Caribou, the logging is unfortunately ongoing.

BEWARE THE LOOKALICHEN

Sometimes, a lichen meal can be an animal's last. Cases of accidental poisoning of cattle, sheep, and Elk (*Cervus elaphus canadensis*) in North America with Tumbleweed Lichen (*Xanthoparmelia chlorochroa*) have made it into the news. Experiments with sheep have shown that usnic acid, a major compound of this lichen, might be the culprit.

Although *Xanthoparmelia* lichens are quite speciose and abundant, they do not usually cause poisonings. Most of these lichens grow closely attached to rock and would be unlikely to attract the attention of grazing animals. Not so the Tumbleweed Lichen: its thallus lies around loosely on the ground, real tumbleweeds in miniature, and grazing animals will not realize they have ingested the lichens until it is too late. And, as shown on page 274, some people have even weaponized lichen substances to poison predators.

Camouflage and mimesis

Camouflage encompasses an astonishing range of spectacular phenomena in the animal world. Animals that resemble a leaf, twig, or flower evade detection by their predators in a never-ending tit-for-tat of adaptation and counter-adaptation. Given that lichen communities form the finely patterned backdrop of this game in many ecosystems, it comes as no surprise that many organisms have evolved to play the lichen card.

For animals living at the lichen scale, the opportunities to camouflage within these colorful communities are endless. Skin patterns serving to blend in with the lichen–covered undergrowth are frequently found in amphibians and reptiles. This even includes lichen communities on leaves, as in the Central American Canal Zone Tree Frog (*Boana rufitela*) or in the Tanzanian Uluguru Forest Tree Frog (*Leptopelis uluguruensis*). Little doubt hangs over the purpose of lichen-moss camouflage of the Madagascan leaf-tailed geckos of the genus *Uroplatus*.

→ Camouflage. The Canal Zone Tree Frog (bottom right) exhibits a simple spotted color pattern, blending in with leaves covered in lichens. Some insects and insect larvae (above, the larva of a moth of the family Psychidae observed in Mexican dry forest) cover themselves with pieces of lichens.

← A master of disguise on lichen- and bryophyte-covered tree bark is the almost invisible Leaf-Tailed Gecko (*Uroplatus sikorae*).

↑   Some Insects seem to be opportunistically overgrown by lichens, such as this neotropical mantid *Choeradodis rhombicollis*.

Camouflage (blending in with the environment) is often difficult to distinguish from mimesis (resembling an inanimate object). The Central American Tailed Wax Bug (*Alaruasa violacea*) resembles common tropical crust lichens in the genus *Arthonia*, including the black thallus borderline. Whether this indeed constitutes mimesis is open to question, as it has also been suggested that the waxy tail mimics attack by a fungus, making the insect unattractive to predators. *Chytonix griseorufa*, a moth of southeastern Brazil, resembles the common and widespread Christmas Lichen (*Herpothallon rubrocinctum*), and it occurs in one of the regions where this lichen is most abundant.

Not to be outdone, some insects *wear* lichens, to help them blend in with their general environment. Caterpillars of the family Psychidae are covered with pieces of lichens, like roof tiles. They disproportionately use species rich in chemical compounds. Long-lived individuals of the neotropical mantid genus *Choeradodis* are colonized by tiny, normally leaf-dwelling lichens because they offer a suitable substrate resembling a leaf. Last but not least, weevils of the genus *Gymnopholus* from Papua New Guinea, including the aptly named Lichen Weevil (*G. lichenifer*), host entire communities of lichens, supported by special hairs and secretions from the weevil. To top it off, the mobile lichens in turn support unique mite communities: an entire food chain in a weevil backpack.

↖↙ Mimesis. The moth *Chytonix griseorufa* appears to mimic a red-bordered Christmas Lichen (*Herpothallon rubrocinctum*), and the bug *Alaruasa violacea* resembles an *Arthonia* species.

→ Grasshoppers and their relatives (Orthoptera) include a number of species with lichen-related camouflage. The spectacular neotropical Maculated Leaf Katydid (*Anapolisia maculosa*—A) features two distinctive spot patterns on its wings, one resembling the lichen genus *Calopadia* and the other the genus *Gyalectidium*, both common in lichen communities on living leaves. *Dissonulichen hebardi* (B) is a perfect match for gray macrolichens of the family Parmeliaceae. The South American usnea katydids of the genera *Apolinaria*, *Lichenodraculus*, *Machima*, *Machimoides*, and *Markia* (here *Markia hystrix*—C) perfectly blend in with beard lichens of the genus *Usnea*, though curiously they are rarely found on these lichens (because we don't see them?).

Home sweet microsite

Lichens are famously finicky about where they grow. For dedicated lichen ecologists, the study of lichens often leads down the path of studying microclimate and the chemistry of different rock and bark types. Some lichens might grow only on one kind of tree, or on rock of specific chemistry, and for these specialist species the spectrum of substrates is narrow. Other species, generalists, may grow on a broad spectrum of natural substrates.

READING THE ENVIRONMENT THROUGH LICHENS

Many lichens have such specific habitat preferences that identification guides use their substrates—rock, bark, soil—as a faster way to identify them than their thallus traits. In fact, a large percentage of all lichens are substrate specialists. Learning lichen species and their requirements is like acquiring a Rosetta stone to read an ecosystem, with each lichen a word, the totality of lichen composition a story of place.

→ High species diversity of pin lichens of the genera *Chaenotheca* (inset *C. chrysocephala*), *Sclerophora*, and *Calicium* is characteristic of old forests. The species are often hyperspecialists, growing on rain-sheltered bark of old trees (here, an old-growth forest in the Seymour River drainage of inland British Columbia). A study in Canada found that species numbers of pin lichens increase continuously with increasing stand age, not leveling off even after four centuries.

While this is true of many organismal groups, lichens write their stories at levels of granularity that go beyond the wider story told by the community of vascular plants or birds. In a typical forest stand, lichen composition not only tells us about the underlying bedrock, climate, humidity, and air quality, but also reveals favorite canine pee spots, popular bird perches, and hints about the stand's history. If you know how to read it.

At the scale that matters for lichens, slight differences in substrate chemistry determine where each species can make a living. Lichen composition is polarized by pH in a way reminiscent of how people are polarized by politics: specific species occur at high pH, others at low pH, and swing voters are rare. In practical terms, this means limestone (high pH) has a lichen composition that shows almost no overlap with that of granite (low pH).

Lichens are also affected by bark chemistry and porosity, and tree bark also exhibits pH differences. In temperate climates, this is easily seen in the differences between the composition of lichens of broadleaved trees and conifers, which exhibit very different bark chemistry. Tree species within these groups likewise exhibit fine-scale differences in lichen composition. Some lichens are not so finicky about where they grow, and show up on all kinds of substrates: they occur on bark of almost any tree, on rock, on decomposing leaves, on bare mineral soil. These generalists often comprise a major part of lichen biomass that people see, and some become weedy, accompanying human life in familiar environments.

↓　*Anzia centrifuga* is listed as Vulnerable on the IUCN Red List of Threatened Species. It is only known from two small populations on a volcano on the island of Porto Santo (Madeira).

OLD FORESTS: RICHNESS THROUGH STRUCTURE?

Lichenologists performing detailed species inventories in Britain in the 1970s began to notice that certain species were found only in very old woodlands. Ever since, lichen studies have documented similar phenomena in regions around the globe, and ecologists have drawn up lists of "old-growth forest indicator species" that apply to their local ecosystems. The lichens involved come from diverse evolutionary groups. Reliable members of the old-growth cast include pin lichens, an evolutionarily diverse group of lichens so named because the fruiting body vaguely resembles a sewing pin. Others include a peculiar group of macrolichens called coral lichens, numerous lichens with cyanobacterial symbionts (cyanolichens), and a large number of crust lichens.

Old-growth forests have a certain allure that blurs into the enchanted forests of folklore, from the 4,000-year-old *Epic of Gilgamesh* (which involved cutting down trees in an old-growth forest) to *Hansel and Gretel* and *Lord of the Rings*. But, enchantment aside, why would a particular lichen be found only in an old forest and not, say, in a medium-aged forest? This question has intrigued many ecologists, and there exist several leading theories, which are not necessarily mutually exclusive. One of the earliest of these holds that old-growth forest lichens require "ecological continuity," in other words little change over time—but this is hard to prove.

More testable hypotheses have been advanced that suggest that as forests age, they acquire specific types of "forest structure" absent in younger forests. This structure includes trees of greater height and larger diameter that begin to have a larger "rain umbrella" effect, with sheltered areas at their base and in their increasingly furrowed bark; the presence of standing dead trees of various decay stages, bark on or off; and dead and dying wood lying on the ground. These structural attributes go hand in hand with more constant and often cooler temperatures than outside the forest, as well as more constant relative humidities.

Once a forest is old enough to have the requisite structure to support the most finicky of species, it is possible that its increasing age alone could play a role in picking up ever more species. This idea, called the "petri dish hypothesis" by the Canadian lichenologist Trevor Goward, suggests that lichen colonization occurs as a series of hit-and-miss inoculations by spores over time. In essence, an old forest is like a lab petri dish left uncovered for centuries; the longer it's there, the greater the chances that the spores of even the rarest species will come to settle in it.

Regardless of how their lichen composition comes to be, old-growth forests are on the shortlist of natural wonders that themselves could be confined to historical folktales, like glaciers and polar bears, if human attitudes to our planet do not change. Old-growth forests continue to be felled for lumber, and many of the rare lichens associated with them are tracked on endangered species lists around the world (page 238).

LICHEN CONSERVATION

Since 1948, the International Union for the Conservation of Nature (IUCN) has been concerned with the protection of wildlife and plant and (much later) fungal species, evaluating their conservation status and placing them on so-called Red Lists. The *IUCN Red List of Threatened Species* has become the primary reference depicting extinction risks of animal, plant, and fungal species. Thus far, over 150,000 species have been assessed, and over 40,000 listed as threatened or extinct. As of 2022, this list includes only 86 lichens and 539 other fungi, and most summary statistics report on only plants and animals.

Environmental monitoring projects using lichens have sparked great interest around the globe in community science projects. They have raised awareness about their biology and diversity and how they can be used to assess environmental quality right on one's doorstep. Such projects are now being fostered, and propelled into the digital age, by increasingly popular community science applications such as *iNaturalist*, which employ artificial intelligence to aid in species identification.

Lichens on artificial substrates

A seeming paradox of the sensitivity and environmental selectivity of many lichens is the appearance of lichens on artificial substrates. Generalist species expand beyond their range of natural substrates to take over walls and buildings, gravestones, street signs, park benches, abandoned cars, clotheslines and telephone cables, plastic, and even old bottles and weathered textiles and shoes. Sometimes, human-made surfaces can provide substrates for specialists in landscapes where their needs are not otherwise met.

The best-known artificial lichen substrate is surely the gravestone. Gravestones come in a variety of shapes, sizes, and materials, including marble and granite. In rock-free landscapes, they provide a habitat for rock-dwelling lichens, and thus contribute substantially to local lichen diversity. Best of all, though (for the nerdy types), gravestones have dates. Assuming they have not been scrubbed clean, gravestones can function as an excellent outdoor laboratory for studying lichen colonization and growth rates.

This combination of attributes led Anne Pringle, a lichen researcher at Harvard University during the early 2000s, to use lichens on gravestones to study whether, as lichens age, they are more or less likely to die. In almost poetic irony, Pringle found that lichen survival rate *increased* with age. If sexual reproduction happens at around 15–20 years of age, as has been inferred for some lichen fungi, this suggests that many more generations of living fungal great-grandparents may overlap with newborns than is the case with us or most animals.

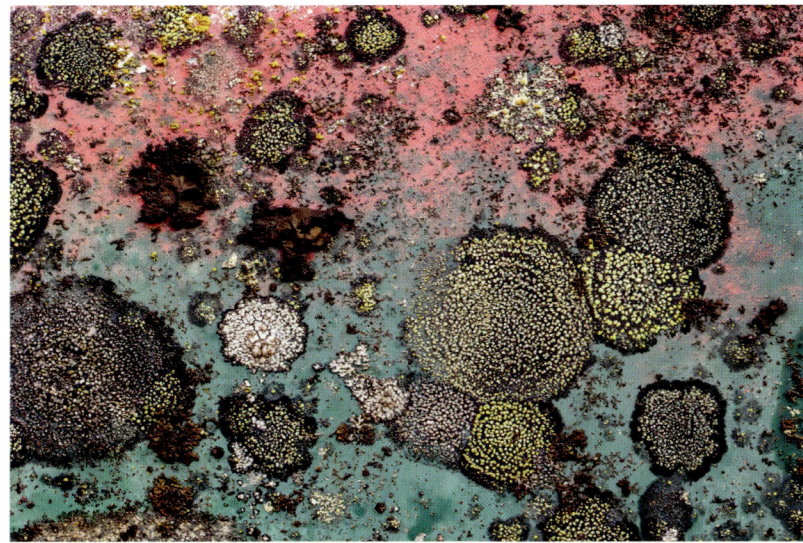

↑→ A long-abandoned truck in British Columbia, Canada, covered by diverse, normally rock-dwelling lichens.

← Resting in peace: lichens on a limestone headstone in Shobrooke, Devon (England).

← Abandoned tires of farm equipment in Puerto Rico supporting sizable thalli of *Coccocarpia* and *Cladonia* lichens.

↓ *Xanthoparmelia mougeotina*, normally a rock-dwelling lichen, found a home on this glass bottle in Australia. *Follmannia orthoclada* developed on a broken piece of glass in the Atacama Desert.

More surprising, perhaps, is the appearance of lichens on substrates less obviously similar to anything in nature. Given the right climatic conditions and a parked car, lichens can become established on metal, paint, glass, and even tires. Many of these species are generalists that in nature might occur on rock, bark, or soil, but cars can even host hyperspecialists. Some lichens occur in nature only on particularly iron-rich rocks—in a landscape poor in such rocks, a piece of parked rusted steel might just fit the bill.

The same goes for a variety of other artificial substrates. The surfaces of old glass bottles can occasionally host lichens found on rock, while abandoned construction textiles, clothes, and shoes provide a weathered, textured surface that quickly gets enriched in microbial biofilms—much like bark—and eventually can host lichens.

LEAF DWELLERS AS LAB RATS

The ability of leaf-dwelling lichens to grow on artificial plastic surfaces has turned out to be convenient for scientists. Using microscope cover slips, researchers observed all stages of the life cycle of these otherwise hard-to-study symbioses, while plastic leaves helped to test the presumed influence of so-called "drip tips," the prolonged tips found in many rainforest leaves, on the development of leaf-dwelling lichen communities. Because they take a much shorter amount of time than most lichens to complete their life cycles, leaf-dwelling lichens are also promising future models for the *in vitro* resynthesis of lichen symbioses.

→ Testing the effect of "drip tips" using plastic leaves takes patience: It required three years of experimental care and waiting before the leaves could be "harvested" and analyzed.

↓ This plastic warning sign at La Selva Biological Station in Costa Rica was found to harbor no fewer than 63 different lichen species.

Urban lichens

More than half of all people live in cities, and this proportion is projected to increase to two-thirds by 2050. Although urban land areas amount to only 775,000 sq miles (2 million sq km) worldwide, they have developed into their own ecosystems with characteristic lichen communities.

Unnoticed by most, lichens are ubiquitous in urban ecosystems, covering both tree bark and human-made materials, such as walls, pavements, metals, and even plastic. The importance of lichens in urban ecosystems was recognized as early as 1866 by the Finnish lichenologist and mycologist William Nylander, who realized that urban landscapes affected the diversity and composition of lichen communities and anticipated the use of lichens as bioindicators of environmental quality. Today, urban habitats count among the best-studied areas in lichen ecology.

Urban lichen communities can be quite rich, even in large metropolitan areas. Their composition varies depending on where in the world you are. A city such as Lima, Peru, situated in a coastal tropical desert, has different lichen communities than Bogotá, Colombia, located high in the humid northern Andes. Both, in turn, differ markedly from temperate cities such as Berlin or Beijing.

→ Perhaps the most common and most familiar urban lichen is the Chewing Gum Lichen widespread in the northern hemisphere. It is common on sidewalks and even highway asphalt.

→ The Yellow Wall Lichen (*Xanthoria parietina*) covers extensive areas on this stone balustrade in the city of Schwerin, Germany.

↓ A familiar sight on urban trees is the Candleflame Lichen (*Candelaria concolor*). It looks much the same in Rio de Janeiro (Brazil, left), Berlin (Germany, middle), and Tsukuba (Japan, right).

Even so, some species are nearly ubiquitous in cities around the globe. One is the Candleflame Lichen (*Candelaria concolor*), which exhibits an almost weedy character. So does the Yellow Wall Lichen (*Xanthoria parietina*), considered to be one of the few lichens that is invasive outside its native range. The Chewing Gum Lichen (*Protoparmeliopsis muralis*) is named for its resemblance to, you guessed it, sidewalk chewing gum, with which it shares its habitat. Chewing gum is evidently not the first thing that comes to mind for everyone who sees this lichen, however. In 2014 in the Bavarian town of Wolnzach, *Protoparmeliopsis*-splotched pavements were reported to authorities by an outraged resident as possible midflight discharge from airplane lavatories. A subsequent investigation determined the splotches to be Chewing Gum Lichens.

The disappearance of lichens in urban settings can have knock-on effects on other organisms. In particular, it has been implicated in the phenomenon of "industrial melanism," the increasing prevalence of dark forms of animals such as moths that rely on camouflage to evade predators. This phenomenon was first described at the turn of the twentieth century. In unpolluted areas, trees were historically covered with lichens, which selected for light-colored moths. In industrial areas where lichens disappeared due to pollution, the tree bark became covered with dark soot, and dark variants of the same moth species became more frequent. The best-studied example is the Peppered Moth (*Biston betularia*), in which under unpolluted conditions the light form has a better chance to escape predation on lichen-covered trees than the dark form.

→ Industrial melanism. Color variation in the Peppered Moth was found to be correlated with changes in lichen communities on tree bark due to pollution.

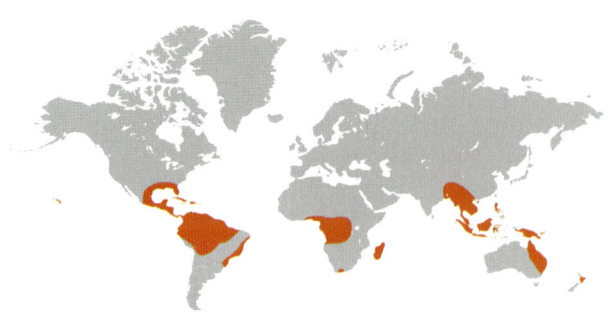

Delicate Leaf-mining Lichen

Sneaking in through the leaf surface

SCIENTIFIC NAME	*Racoplaca subtilissima* Fée
PHYLUM, FAMILY	Ascomycota, Strigulaceae
GROWTH FORM	Crust lichen with delicate, radiating lobes
SPECIES IN GENUS	7
HABITAT	Plant leaves in tropical rainforests
NOTABLE FEATURES	Grows beneath the leaf cuticle

Lichens growing on living leaves are not necessarily a familiar sight, unless one lives in tropical or subtropical regions. Yet over 1,000 different species of lichen fungi are found growing on this unusual substrate. One may wonder what these tiny lichens are doing to the leaves—which, after all, are the carbon factories of the supporting plants. Well, usually nothing.

Leaf-dwelling lichens do not penetrate the leaf, and only grow on the surface. They also do not interfere much with the leaves' photosynthesis, although they presumably take away some of the incoming light. Lichens usually cover only a small part of the leaf, and they develop richer communities only on relatively old leaves, "old" meaning 3–5 years. At that age, leaves of tropical plants have changed from carbon producers to waste-disposal facilities. So it seems that a very subtle balance has evolved so that leaf-dwelling lichens do not put their supporting plants at a disadvantage.

Well, there is perhaps one exception. Lichens of the closely related genera *Puiggariella*, *Raciborskiella*, *Racoplaca*, and *Strigula* grow beneath the leaf cuticle. They do so because their photobiont is the genus *Cephaleuros*, in the free-living state a half-parasite on leaves. The lichen often grows along the midrib or around holes in the leaf, its fungus apparently using these as entry points to take over a previously non-lichenized *Cephaleuros* alga.

→   *Racoplaca subtilissima* forms minute thalli with delicate, elegantly radiating lobes on the surface of living leaves. Next to it is a non-lichenized *Cephaleuros* alga, also the photobiont in the lichen.

PANNARIA ANDINA

Andine Shingle Lichen

Resting on a sponge

SCIENTIFIC NAME	*Pannaria andina* P. M. Jørg. & Sipman
PHYLUM, FAMILY	Ascomycota, Pannariaceae
GROWTH FORM	Neatly branched foliose lichen with numerous apothecia
SPECIES IN GENUS	40
HABITAT	On shrubs in wet paramos in the northern Andes
NOTABLE FEATURES	The thallus rests on a thick, spongy, black hypothallus

Pannaria andina has what it takes to survive in an ecosystem that has been described as "summer every day, winter every night."

The paramos of South America are wet, upper montane grasslands and shrublands developed above the treeline, at over 11,500 ft (3,500 m) altitude. They look about the same all year long, but day and night make a big difference. Temperatures can reach 86 °F (30 °C) in plain sunlight, and easily drop to below freezing at night. Precipitation may come as rain, but also as cloud cover, and nutrients are sparse at high altitudes.

To overcome the scarcity of nutrients, *Pannaria andina* associates with *Nostoc* cyanobacteria as its primary photobiont, capable of fixing atmospheric nitrogen (page 200). Hence the bluish-gray color of the lichen, strongly contrasting with a black mat of densely entangled fungal hyphae that supports the thallus as a so-called hypothallus. What looks like an air mattress giving the lichen comfort is likely a water-catching device: the fine and dense net of hyphae combs water through condensation out of the humid air or cloud cover. It might also help to ventilate excessive water.

Although conspicuous and not uncommon in its range, *Pannaria andina* was only described by Western science two decades ago. It is found from the high mountains of Costa Rica through the Andes to Chile.

→   *Pannaria andina* adorning a branch of a shrub in the paramo of Sumapaz, the largest paramo in the world, right outside Colombia's capital Bogotá.

CARBONEA VORTICOSA

Abominable Coal Dot

Peak bagger

SCIENTIFIC NAME	*Carbonea vorticosa* (Flörke) Hertel
PHYLUM, FAMILY	Ascomycota, Lecanoraceae
GROWTH FORM	A crust lichen that mostly lacks a visible thallus and is recognized by its coal-black apothecia
SPECIES IN GENUS	20
HABITAT	On rocks in mountain regions
NOTABLE FEATURES	Goes where few lichens have gone before

Lichens in the genus *Carbonea* are for connoisseurs. They are part of a species-rich group that grow on rock, in which all of the most important information for their identification, like spore shape, ascus features, and other anatomical detail, is contained inside the apothecia and needs to be examined with high-power microscopy. But they have a fan club, and reward the well-trained eye with records from some of the most extreme habitats on Earth.

Most coal dot lichens do not possess any visible thallus, but this does not mean they do not have one. The thallus, such as it is, consists of a thin network of fungal hyphae interacting with algal cells in the microscopic air spaces between crystals of otherwise almost solid rock.

Taking refuge in the rock suggests a kind of toughness, and *Carbonea vorticosa* has the statistics to back it up: it holds the distinction of the highest-known record of a lichen on Earth. It has been collected from an astonishing 24,000 ft (7,400 m) elevation on Makalu in the Himalayas, the fifth-highest mountain in the world, an elevation 3,300 ft (1,000 m) higher than the highest-recorded moss and almost 4,300 ft (1,310 m) higher than the highest-recorded vascular plant.

→ Apothecia of *Carbonea vorticosa* emerging from cracks of quartzitic rock. Wondering where the thallus is? It is *inside* the rock, lining the narrowest cracks between the quartzite crystals.

Western Waterfan

Aquatic lichen lifestyle

SCIENTIFIC NAME	*Peltigera gowardii* Lendemer & H. E. O'Brien
PHYLUM, FAMILY	Ascomycota, Peltigeraceae
GROWTH FORM	Foliose lichen with fan-shaped, radiating lobes
SPECIES IN GENUS	100
HABITAT	Attached to rocks in cool freshwater streams
NOTABLE FEATURES	One of the few fully aquatic macrolichens

Lichens involving fungi of the genus *Peltigera*, often called pelts, are a common feature of temperate ecosystems and commonly live on the ground. One of them, *Peltigera polydactylon*, is a well-known model species in which the exchange of goods and services in lichen symbiosis was first studied. Like most lichens, they require a regular regime of drying and die from prolonged waterlogging. Not so the Watershield and the Western Waterfan—two species in the genus that have transitioned into a fully aquatic lifestyle.

The Watershield, *Peltigera hydrothyria*, was discovered for science on July 16, 1851, by two New England naturalists out on a ramble on Bald Mountain, Vermont, USA. John Lewis Russell, who gave the lichen both its first scientific name and its English name Watershield, described how:

In the company of my friend C. C. Frost … we struck on our descent upon a brook, through whose bed we wended our way, hoping that the water might prove more auspicious than the dry, sunburnt rocks. Nor were we disappointed as the result proved. Our delight was as mutual as it was unexpected, when I drew from the stream an aquatic lichenose plant diffusing a grateful aroma not unlike that of the bark of the black birch, and covered with apothecia, every part in perfect condition.

The source of the aroma, like the various specific aromas of other lichens, remains virtually unstudied, but likely derives from volatile compounds produced by the fungus.

Watershields like their water pristine, but streams that provide the right combination of water quality and consistent flow have grown rare. Recent molecular studies have shown that Watershields in western and eastern North America are distinct, the western population now being named Western Waterfan. Both the Watershield and the Western Waterfan are listed as Threatened in Canada and as Sensitive in the western United States.

→ Western Waterfan, *Peltigera gowardii*, one of two emblematic, fully aquatic species of the genus *Peltigera*.

Cottonwood Glasswhiskers Lichen

Tiny hallmarks of the oldest trees

SCIENTIFIC NAME	*Sclerophora amabilis* (Tibell) Tibell
PHYLUM, FAMILY	Ascomycota, Coniocybaceae
GROWTH FORM	Crustose lichen with apothecia on stalks
SPECIES IN GENUS	6
HABITAT	Old-growth mixed conifer and hardwood forests of the upper temperate and boreal zones
NOTABLE FEATURES	Part of a group of species found only in old, unlogged forests

Sclerophora amabilis is part of a large group of species from different genera and families and different evolutionary origins. They are characterized by fruiting bodies borne on stalks. Because of this unmistakable shape, they are also called pin lichens, or calicioids, in reference to the genus *Calicium* where many of them have been classified.

Different species of calicioids are characterized by color, spore shape, and symbiotic association (which alga the fungus consorts with). Their special habitat requirements have led to them being closely studied by forest conservationists.

Like most pin lichens, *Sclerophora amabilis* requires a very specific combination of environments. First, the forests in which it lives are usually very old, with towering trees and wood in all different stages of decay. Second, the tree on which the lichen grows must have just the right bark conditions so that its pin-like fruiting bodies are protected from physical damage. The pins are fragile, so they grow in bark furrows that protect them from abrasion by sliding snow and ice, and often they appear concentrated in sheltered pockets where they are even protected from direct rainfall.

Pin lichens have become increasingly rare, and are threatened by the onward march of forestry. Forest practices around the world are most economical when large tracts of land can be treated the same way, and while stands are often replanted, it can take centuries for a forest to again achieve the characteristics needed by lichens such as the Cottonwood Glasswhiskers. Accordingly, various pin lichens feature on the Threatened and Endangered Species lists of numerous countries.

→ *Sclerophora amabilis* is easily recognized even with a hand lens, based on the light beige color of its "heads" and the often yellowish color of its stalks. In general, however, the identification of pin lichens requires microscopic examination.

Ancient Coral Lichen

Denizens of the primeval forests

SCIENTIFIC NAME	*Sphaerophorus venerabilis* Wedin *et al.*
PHYLUM, FAMILY	Ascomycota, Sphaerophoraceae
GROWTH FORM	Gray fruticose thalli with round branches and downward-pointing apothecia at their tips bursting into masses of dry spores
SPECIES IN GENUS	6
HABITAT	On bark and wood of conifers
NOTABLE FEATURES	Found only in ancient forests

Coral lichens are part of an unusual group of lichen symbioses including the genera *Sphaerophorus*, *Austropeltum*, *Bunodophoron*, *Calycidium*, *Leifidium*, and *Neophyllis*, most of which form easily recognizable, fruticose or irregularly flattened thalli that recall the polyps of coral reefs. Their apothecia look like those of pin lichens, producing masses of dry spores. Coral lichens are especially speciose in temperate rainforest regions, though a few species extend into Arctic and alpine tundra, as well as into tropical mountain ecosystems.

The coral lichen symbiosis is unusual in many ways. The six genera mentioned above appear to represent a single origin of macrolichen morphology separate from other macrolichens, with high diversity and endemism in parts of the southern hemisphere that were contiguous during the time of the supercontinent Gondwana, up until the Paleogene (about 66 to 23 million years ago).

The algal symbiont in *Sphaerophorus* has been found to be a species of *Dictyochloropsis*, more familiar as the main green algal partner in numerous tripartite lichens (page 134). Many species of *Sphaerophorus* and the closely related *Bunodophoron* display a striking obligate dependence on old-growth forests. The underlying reasons for this are unknown, but whatever they are, they appear to leave the lichens little wiggle room. Few species of coral lichens are found in young forests or disturbed habitats, even when plenty of their propagules are available nearby.

Although coral lichens are not classified as endangered in most jurisdictions, this might soon warrant reconsideration. The big, old trees in forests such as those in which the Ancient Coral occurs are often hollow from heart-rot fungus, a natural state of affairs that can go on for centuries and is part and parcel of a healthy ecosystem. To this day, such forests continue to be felled and replanted with fast-growing timber—posing a major threat to coral lichens and the many less visible old-growth-dependent organisms that occur alongside them.

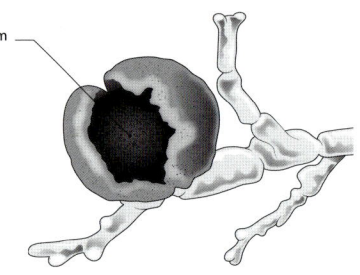

Mazaedium

Dispersed by wind

Sphaerophorus venerabilis produces characteristic fruiting bodies in which the spores accumulate as a blackish mass, so-called mazaedia, which are then dispersed by wind.

→ *Sphaerophorus venerabilis* in the still unprotected Rainbow-Jordan Wilderness in British Columbia, Canada. Ancient Coral Lichen is never found in young forests, and is threatened by unregulated forestry practices.

ERIODERMA PEDICELLATUM

Boreal Felt Lichen

Flagship of lichen conservation

SCIENTIFIC NAME	*Erioderma pedicellatum* (Hue) P. M. Jørg.
PHYLUM, FAMILY	Ascomycota, Pannariaceae
GROWTH FORM	Foliose lichen with distinctly felty upper and lower surfaces
SPECIES IN GENUS	40
HABITAT	On trunks and branches of conifer trees in boreal rainforests
NOTABLE FEATURES	"Umbrella species" of highly endangered forest habitats

While most lichens are in one way or another finicky about where they grow, felt lichens are *very* finicky. As in all cyanolichens, their photobiont requires liquid water to photosynthesize, but unlike many, felt lichens don't seem to tolerate low humidities. *Erioderma* species are mainly found in cool high-elevation forests in the tropics. A few species have, however, made their way north to that most threatened of ecosystems, the boreal rainforest.

Erioderma pedicellatum was described as a new species based on a specimen collected in 1902 in New Brunswick, Canada. Thirty-six years later, it was discovered across the ocean in the boreal rainforests of Norway, and again described as new, this time as *Erioderma boreale*. Its rarity was immediately recognized, and the forest where it was found was set aside as a nature reserve in 1952. However, the reserve proved too small, and logging of the surrounding forest dried out the lichen's habitat; it disappeared. The species was not seen again

in Europe until 1994, and by then taxonomists had realized that it was the same species as the one in Canada.

Today, traditional land guardians of the Miawpukek First Nation in Newfoundland teach schoolchildren about the lichen, which they credit with helping save parts of the Bay D'Espoir rainforest from logging. It is also what conservationists call an "umbrella species": its recognition and high profile have helped achieve conservation for the hundreds of other, lesser-known species that also require these special habitats.

The Boreal Felt Lichen was the first lichen to be placed on the Red List of the International Union for the Conservation of Nature (IUCN) in 2003, and achieved national rank under Canada's Species at Risk Act in 2005. It has since been found in boreal rainforests of Alaska and the Kamchatka Peninsula of Russia, but because of its sparse, disconnected populations and the fact that the forests it occurs in yield valuable timber, it remains one of the most threatened lichens in the global north.

→   The Boreal Felt Lichen (*Erioderma pedicellatum*) has become a poster species for the global efforts to protect threatened lichens and their habitats.

LICHENS
AND PEOPLE

Medical promise

Centuries before scientific naming conventions and lab bench experiments, a tradition of medical uses prompted early research into the natural history of lichens across many cultures. Though methods and standards of proof have changed, the interest in pharmaceutical applications derived from lichens has carried on unabated into the present day.

LICHENS IN HERBAL AND TRADITIONAL MEDICINE

The Lung Lichen (*Lobaria pulmonaria*) might be the first lichen depicted in a medieval herbal, the *Gart der Gesundheit* ("Garden of health") from 1485, as "Pulmonaria." Its name—like that of the confusingly similar-sounding plant, the Lungwort (*Pulmonaria officinalis*)—derives from the "Doctrine of Signatures," a concept going back to Greek physician and pharmacologist Dioscorides and further developed by the Swiss physician and alchemist Theophrastus von Hohenheim (1493–1541), better known as Paracelsus.

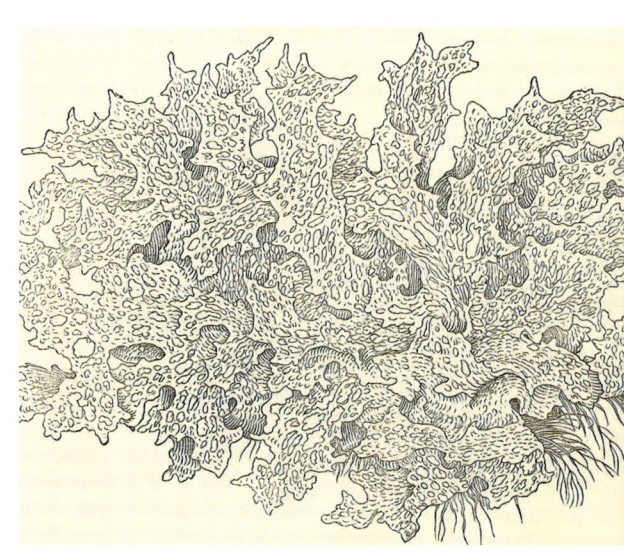

According to Paracelsus, "Nature marks each growth … according to its curative benefit"—in other words, plants that resemble parts of the body could be used to cure illnesses of that part. The Lung Lichen, you guessed it, got its name from a fancied resemblance to the interior of a lung.

For centuries, cultures around the world have been using lichens in medicine and in rituals, but few such applications are documented. The recently described basidiolichen *Dictyonema huaorani* from Amazonian Ecuador has been used in rituals of the Huaorani; it was found to potentially contain psilocybin, a hallucinogenic substance known from *Psilocybe* mushrooms. The indigenous Dení people from the western Brazilian Amazon region use the yellow powder of the medulla of a crust lichen (possibly *Astrothelium aeneum* or *Pyrenula ochraceoflava*) for sniffing.

A curiosity of medicinal uses in Europe in the late medieval period are the so-called "skull lichens." At the time, human remains in the landscape were probably a common sight, and bleached skulls became overgrown with lichens and mosses. This led to the belief that these lichens, mostly from the genera *Hypogymnia*, *Parmelia*, and *Usnea*, were beneficial in treating diseases. Paracelsus was apparently the first to document the use of lichens growing on skulls to treat blood clotting, mentioning the "*Mieß auff den todten Köpffen*" ("The moss on the dead skulls") in the tenth chapter of his work *Grosse Wundartzney* from 1536.

←← The Lung Lichen (*Lobaria pulmonaria*) over four centuries. The *Gart der Gesundheit*, first published in 1485 and reprinted in several editions (here, the 1487 edition), was the first European herbal to depict the Lung Lichen, though in a highly stylized way.

←↑ In his *De Historia Stirpium* of 1542, Leonard Fuchs gave a more realistic rendering, topped by a copper engraving more than 200 years later, in the *Botanica Pharmaceutica* from 1791.

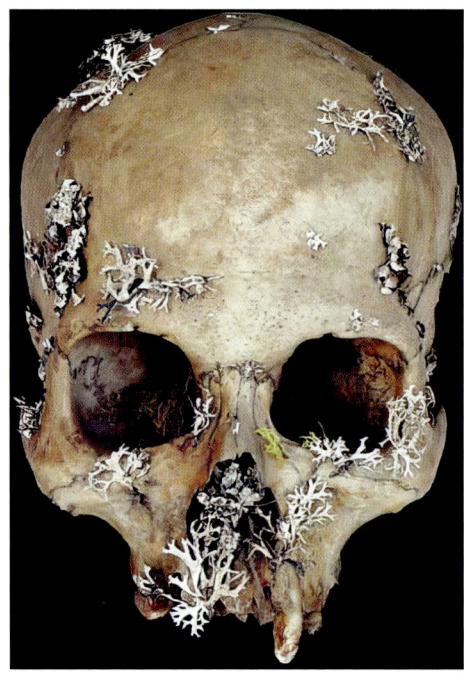

Usnea cranii humani.
Hirnschedel - Mooß.

LICHEN MEDICINALS

The use of lichens in traditional medicine of non-European cultures was based on empirical research and experience, more accurate than the pseudoscientific Doctrine of Signatures, and it brought to light potential applications that were later found to derive from the lichens' complex biochemistry. Many bacteria that grow on lichens possess enzymes that degrade lichen symbionts, and fungi have evolved a plethora of substances that inhibit their growth, acting as antibiotics.

The most widely used antibiotic originating from lichens, usnic acid, was obtained from conspicuous and easy-to-collect macrolichens, such as *Usnea*, *Alectoria*, and *Ramalina*. Lichen polysaccharides have expectorant properties, and have been used in treating coughs and other respiratory illnesses. Today, lichen extracts continue to be used as the basis for natural-product medicines, mostly from *Cetraria islandica* (page 266), *Lobaria pulmonaria*, and species of *Usnea*. *Xanthoparmelia scabrosa* powder has even been suggested to enhance male stamina, although evidence for this is hard to come by.

Hundreds of studies on antioxidant, antibacterial, antiviral, anticancer, and other effects of lichen extracts and substances have been published, but few have made it into drug trials. Unlike other fungi, no lichen appears to have been the source of a commercially produced drug beyond natural medicines, neither pure substances nor synthetic drugs derived from lichen precursors. However, there are promising studies looking at the HIV-inhibitory properties of *Umbilicaria esculenta*, and

← Skull lichens, a late medieval curiosity named *Usnea cranii humani* that appeared along with the beginning of the systematic documentation of lichens in European medicine, here depicted from the *Theatrum Botanicum* from 1744, beneath an actual skull overgrown with lichens.

↗→ Bless you! Tropical crust lichens with intense yellow-orange pigments (anthraquinones) were used by the Amazonian Dení people for sniffing, causing tingling and sneezing (which by many is considered pleasant). Common and widespread rainforest species with such pigments are *Pyrenula ochraceoflava* (left) and *Astrothelium aeneum* (right).

at a possible reduction of the effects of Alzheimer's disease by orcein, a precursor of lichen dyes (page 250).

The greatest obstacle to using lichens on an industrial scale is their slow growth and the difficulty of recreating them *in vitro*. To obtain 1 g of useful substance, about 100 g of lichen is needed, making collection in the wild unsustainable. Also, while many lichen fungi can be cultured, their biochemistry is not necessarily the same as in symbiosis. However, the use of mycobiont cells immobilized in a specific matrix to maintain their enzymatic activity, explored in studies by the Brazilian lichenologist and chemist Eugênia Pereira, is a promising tool for potential industrial applications.

Lichen dishes

Fungi have been a part of the human diet since at least the Stone Age, and probably long before that. Baker's and brewer's yeasts (*Saccharomyces*) as well as blue cheese molds (*Penicillium*) are essential for the production of bread, cheeses, and alcoholic beverages. The Mexican delicacy *huitlacoche* is derived from the parasitic fungus *Ustilago maydis* on corn, and mushrooms feature in many cuisines. But what about lichens?

Whether we realize it or not, most of us eat fungal-derived products on a daily basis. The cliché wine, bread, and cheese picnic would not be possible without fungi; neither would consumption of any grocery-store product containing citric acid. But who consumes lichens? Although lichens are ubiquitous, they only form a consistent element of the human diet in certain cultures. Few lichens are known to be edible and nutritious, and most contain polysaccharides that are difficult to digest. Moreover, their slow growth makes large-scale consumption unsustainable.

In cultures in which lichens do appear on the dinner plate, they seem to have some staying power. The Rock Tripe (*Umbilicaria esculenta*) is native to East Asia and known as *iwatake* (or *wawatake*) in Japan, *shi'er* in China, and *seogi* in South Korea. It is considered a delicacy, consumed in the form of salad, soup, or fried chips. It may have first been mentioned in 1643 in the Japanese cookbook *Ryōri Monogatari*, which lists 15 edible mushrooms including lichens. First Nations across the northern half of North America have eaten relatives of the Asian Rock Tripe in various ways, including as chips. The Cree added it to fish broth to nourish the sick, although the neighboring Inuit considered a prolonged diet based exclusively on these lichens potentially harmful (suggesting that it was tried).

Chinese cuisine includes various species of *Lobaria* and *Sticta* in soups and salads. *Parmotrema perlatum*, called Black Stone Flower, or *dagad phool/pattar ke phool* in Hindi, is used in many cuisines in southern India and features in the garam masala and Malvani masala spice mixtures popular in many curries.

Contemporary cuisine has rediscovered lichen delicacies, particularly reindeer lichens of the genus *Cladonia*. The famous Noma restaurant in Copenhagen, briefly ranked among the 50 best restaurants worldwide but now closed, was known among other things for its unique lichen dishes.

↗ Contemporary lichen delicacies: lung lichen omelette served in China (top left), Indian lichen curry from *Parmotrema perlatum*, also known as Black Stone Flower (top right), and a lichen dish formerly offered by the famous Noma restaurant in Copenhagen, Denmark (bottom).

Lichens for fun and profit

In addition to medical applications, the industrial use of fungi has become a multi-billion-dollar business, encompassing everything from food resources to bioremediation. But also traditional uses of lichens are experiencing a renaissance.

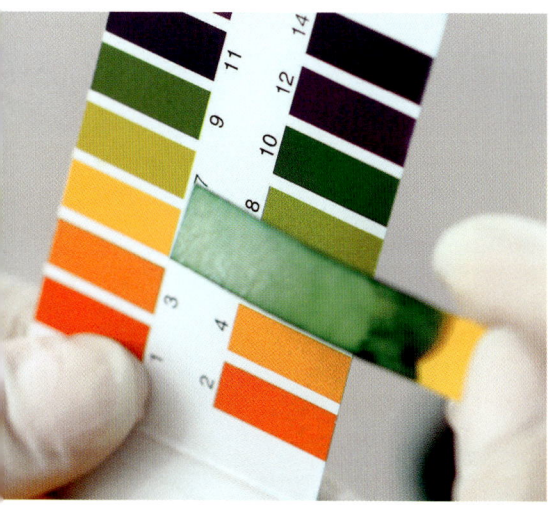

pH, PERFUMES, AND PINTS

High-school chemistry would hardly be complete without the litmus test, the simplest and cheapest way to determine the acidity or alkalinity of a solution. If the solution is acidic, the litmus paper turns red; if alkaline, it turns blue or purple. Few are aware that litmus paper is based on lichen extracts containing derivatives of orsellinic acid, mostly from the lichen historically called *Roccella tinctoria* (now *Roccella phycopsis*). The litmus test goes back to the thirteenth century and lichen extracts were used as early as the sixteenth century. Litmus remains a widespread, basic-purpose pH indicator test in classrooms and laboratories. The "litmus test" has also worked its way into popular culture, among other things as a metaphor in the context of evaluating candidates for potential judicial positions. How the candidate responds to a litmus-test question determines who on the political spectrum will support the nomination (parties that in some countries identify as blue, purple, or red).

Lichens have long been used in the perfumes and cosmetics industry. The best-known lichens in this context are *Evernia prunastri*, Oak Moss, and *Pseudevernia furfuracea*, Tree Moss (neither of which, despite their common names, is a moss). Lichen extracts have been used as fixatives, and to give perfumes an earthy touch, and some well-known perfumes, such as Dior's Eau Sauvage, Eau de Rochas by Rochas, Ralph Lauren's Polo, and Burberry's London for Men, contain Oak Moss. Since 2017, however, the European Union no longer allows lichen extracts in perfumes and cosmetics, because of allergy issues.

Traditional tanning in North America and India applied lichens when converting animal skins into leather. The process used acidic

↑　The Oak Moss (*Evernia prunastri*) was long used in the perfumes industry as a fixative. Numerous brands used to contain lichen extracts in their perfumes, and in some the practice lives on in the name.

←　Litmus test: *Roccella phycopsis* (top) continues to be a source for litmus paper (bottom), used as a basic test to determine the acidity or alkalinity of a solution.

phenolic polymers (tannins), occurring in a broad range of plants and lichens, to bind to collagen proteins in the skin, making it resistant to bacterial attack.

Lobaria pulmonaria, *Cetraria islandica*, and various *Usnea* species have been employed as a supplement for hops or as a source for alcohol fermentation based on lichen starch (lichenin), for brewing beer or making brandy, particularly in northern Europe and Siberia. Modern breweries have rediscovered lichens, and now use them in the production of craft beers such as "Wunderkammer Untappd" and "Goodlife Wildland."

LICHENS TO DYE FOR

Lichens have evolved a broad color palette, ranging from yellow to orange to red, from gray to brown to black, and from light green to bluish or pink to purplish. Some lichen pigments are directly used as dyes, extracted through boiling in water. Examples include the chartreuse highlighter lichens of the genus *Letharia*, containing vulpinic acid (page 274). However, most lichen dyes are derived from a lichen substance that is extracted as a pale brown, crystalline powder: orsellinic acid.

Orsellinic acid is a precursor to various other lichen substances, such as gyrophoric and lecanoric acid, which have a broad range of biological properties (page 68). It also gives orcinol, which through treatment with ammonia reacts to produce variously colored pigments in the range from pink to blue. Lichens containing orsellinic acid or its derivatives react red when a droplet of sodium hypochlorite (better known as bleach) is applied to them (lichenologists call this the "C test"). Besides the aforementioned *Roccella phycopsis*, many other lichens contain these compounds, among them *Dendrographa leucophaea*, *Lasallia pustulata*, *Ochrolechia tartarea*, and *Parmotrema tinctorum*, all of which have been used to produce dyes. Extraction of the substance and production of the various pigments requires prolonged soaking of the lichens in ammonia water. "Prolonged" in this case can mean months, but aficionados of lichen dyeing agree that the results are worth it.

↖↗ Color splash: the centuries-old process of extracting pink to blue lichen dyes from lichens that do not necessarily display those colors in life, skillfully recreated by Emily Donavan and Alissa Allen. They use lichens such as the dried *Umbilicaria* thalli to the left.

→ Chilkat dancing robe. The yellow and blue colors of these robes, used by the Tlingit people along the northwestern coast of North America (Alaska, British Columbia), have been traditionally obtained from yellow to chartreuse highlighter lichens and various lichens containing orsellinic acid derivatives, such as *Umbilicaria*.

Decorative lichens

Vegetation in miniature landscapes is modeled with what is commercially often sold as "Icelandic Moss" or "Ball Moss," but which are actually reindeer lichens of the genus *Cladonia*, mostly *Cladonia stellaris* (the true Icelandic Moss, *Cetraria islandica*, is not used for this purpose). The use of these lichens is not limited to making tiny bushes and trees, but has grown to an industrial scale.

Reindeer lichens also have a long tradition of being used in wreaths and other decorative applications of interior design. A more recent use of reindeer lichens is in so-called "moss walls." Besides their arresting appearance, these have been ascribed numerous benefits, including improved air quality, dust and noise reduction, lower energy costs, and reduced carbon dioxide levels. They are also said to improve productivity, sharpen focus, and reduce stress in workers. All of this could be true if moss walls were made of living plants. Some are, but much of what is commonly marketed as "moss wall" is fabricated using reindeer lichens, dyed in different shades of green and glued to a board.

When it is dyed lichen that is being sold, what is advertised as a "living and breathing moss" wall is little more than a scam. The lichens are neither living nor breathing, nor are they moss, and probably they are accumulating more dust than you would otherwise have. That is not to say that a well-crafted reindeer-lichen wall might not be visually stunning. That alone could brighten and enhance the space of the people working and living around it. But since large quantities of lichens are required for these walls, serious questions about sustainability arise.

The historic Speicherstadt district in Hamburg, Germany, is a UNESCO World Heritage Site. In addition to the impressive Elbphilharmonie building on the banks of the Elbe River, it harbors the world's largest model train display, *Miniatur Wunderland*. The 16,000 sq ft (1,500 sq m) display contains more than 1,000 trains on almost 10 miles (16 km) of track, representing cities and landscapes from all around the world.

Among the 130,000 trees and countless shrubs, there are hardly any examples of what is elsewhere a familiar sight in model landscapes: reindeer lichens. *Miniatur Wunderland* had its trees and bushes carefully crafted, but many model landscapers continue to make use of *Cladonia stellaris* and similar lichens, available in craft supply stores, dyed in various colors, often misleadingly sold as "Icelandic Moss." The fact that these lichens are collected in the wild is worrisome, though companies selling them offer assurances that they are sustainably harvested.

↑ A "moss wall"—reminiscent of a miniature rainforest canopy—is actually composed of reindeer lichens dyed in bright green colors.

← *Miniatur Wunderland*, the world's largest model train display, barely uses reindeer lichens to represent bushes and trees, but elsewhere *Cladonia stellaris* and its relatives continue to be popular in miniature landscape modeling.

Lichens and monuments

When at Machu Picchu, that iconic ruin of the ancient world, most people admire the skill of the builders and take in the breathtaking landscape. Lichenologists have also been known to pull out their hand lenses and scrutinize the diversity of lichens growing on the ancient stone walls. But these natural veneers raise serious concerns among those charged with preserving ancient human heritage.

Human-made constructions overgrown by lichens are a peculiar sight, a miniature replica of a jungle-overgrown ancient temple as in Lara Croft's *Tomb Raider* or its real-world counterpart, Ta Prohm in Cambodia. For some a beautiful sight, and considered ecosystems in their own right, biofilms (surface cover made of small plants, algae, lichens, fungi, and bacteria), including those dominated by lichens, are seen as a problem by others. Many lichens engage in biochemical weathering and calcium oxalate deposition, a property useful in primary succession of rock surfaces and subsequent soil formation, but distinctly unwelcome in the context of ancient stonework and masonry.

Building conservators have long been concerned about how to deal with this issue. One consideration is that the physical or chemical removal of the lichens might do more damage to the surface than if they were allowed to stay. Some studies even suggest that biofilms, in particular endolithic lichens and fungi (those growing inside the rock) may help to protect building surfaces, in much the same way as plant roots may prevent soil erosion and landslides.

Biodeterioration of monuments and buildings by lichens has been studied in several cultural heritage sites, such as the Maya temples in Chichén Itzá in Mexico, the Druid stones in Scotland, the Florentine mosaics in Italy,

the tomb of Cyrus the Great in Iran, and the famous Angkor Wat temple in Cambodia. Measures to remove lichen biofilms or prevent their growth range from chemical biocides, such as household bleach and pentachlorophenol, to presumed eco-friendly nature-based products, such as plant extracts from *Rumex hastatus*, to mechanical cleaning, heat shock treatment, and silicone water repellent.

The jury is still out on which of these is most effective while having the least environmental impact, or whether biofilms should be removed at all. Given the diversity of materials used in buildings and monuments, their age, and the climatic conditions and weathering agents present at a particular site, each case requires individual consideration. Perhaps novel techniques can help keep lichens on monuments while reducing their potential damage to a minimum.

↑   Rock-cut church of Saint George in Lalibela, Ethiopia, overrun by the Yellow Wall Lichen (*Xanthoria parietina*).

←   Machu Picchu in Peru, the "Lost City of the Incas," UNESCO World Heritage Site, constructed in the fifteenth century at almost 8,000 ft (2,440 m) altitude, also harbors an impressive diversity of lichens on its ancient walls. Scientific studies have tried to assess how to deal with them.

Lichens in art

The lichen forms we see today are the result of millions of years of natural selection. They have been called "art forms of nature" (page 110)—though for the symbionts that live in them, of course, they are strictly utilitarian. However, the diversity of their colors and shapes, as well as their intriguing biology, has inspired artists across the centuries.

Lichen art is as diverse as its objects. It can be realistic, depicting lichens in their natural beauty using diverse techniques, in different sizes, such as Kathrin Schlegel's *Feeding from the Tree of Knowledge*. It can be a tiny, stylized replica used in exquisite jewelry. It can be inspired by science and a metaphor for human behavior, such as Kristen Morsches' delicate sculptures *Hidden in Plain Sight*. It can make a statement, using lichens in street art to advocate for clean air, such as Bryony Ella's *Follow the Lichen*. It can be a metaphor for a vision of urban ecology and human relations with nature in cities, such as Oscar Furbacken's wall installation *Urban Lichen*.

Artists are drawn to lichens. The intriguing colors and shapes, the unusual symmetry or lack thereof, the fact that lichens are subconsciously known to many but understood by few, their unique biology— all invite abstraction. Lichen art is ubiquitous.

Lichens have even been featured in mainstream movies, such as in a dream sequence in *The Incredible Hulk*. And the character Sheldon riffed on them in the popular television series *The Big Bang Theory*:

The best organism for human beings to merge with is the lichen itself. That way, you'd be human, fungus, and algae. Triple treat. Like three-bean salad ... picture this, a beautiful outdoor concert. Now, as a human, I appreciate Beethoven. As a fungus, I have a terrific view, growing out of a towering maple tree. And no thank you, expensive concessions. Because as an alga, I'll just snack on this sunlight.

Not least, lichens are frequently used to blend art, community science, and outreach, as in Penelope Cain's *Learning from Lichen*, or *Being Lichen* by artist and lichenologist Nastassja Noell. But enough riffing of our own: let the artwork speak for itself.

↗ *Follow the Lichen*. A powerful street art statement for clean air and zero emissions by UK artist Bryony Ella.

→ *Feeding from the Tree of Knowledge*. This huge *Evernia prunastri*, sculpted by Kathrin Schlegel and standing over 10 ft (3 m) tall, adorns the central campus of the Norwegian University of Life Sciences in Ås, near Oslo (2021; commissioned by Koro Norway/NMBU University of Life Sciences Norway).

← *Urban Lichen.* Oscar Furbacken's larger-than-life wall installation of a photoprint of *Xanthoria parietina* compares the structure of lichens with the layout of a city.

→ Lichen jewelry: hand-made statement pieces by Eileen O'Shea.

↓ *Hidden in Plain Sight.* Sculptures by Kristen Morsches depict lichens interfered with by parasites, a metaphor for human intrusion into nature. The artwork recalls the research of lichen ecologist James Lawrey, a relative of the artist.

↙ Lichen crocheting by Japanese artist Kozue Mihashi.

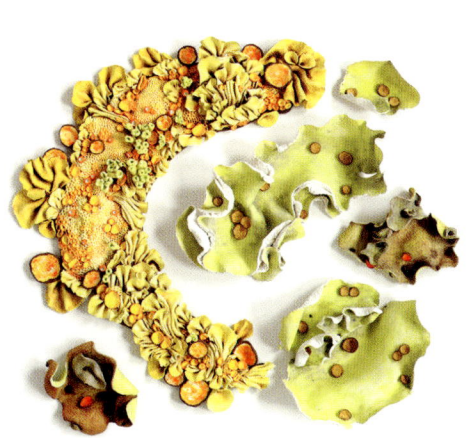

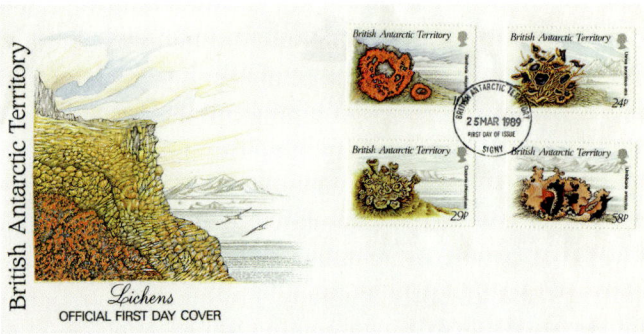

← ↓ A series of lichen postage stamps commemorating the British Antarctic Territory, issued in 1989. This Finnish 2-euro coin from 2022 (inscribed "climate research") depicts a beard lichen.

↙ Lichen jewel. Thomas Barlow's artistic photography highlights the exquisite nature of small pebbles from the Namib desert covered with diverse lichens only known from that region.

→ Painting of the Australasian lichen *Rexiella fuliginosa* by artist Anna Voytsekhovich.

Metaphor: are we all (queer) lichens?

Lichens are the quintessential symbiosis: a stable coexistence of organisms belonging to different domains and kingdoms, resulting in a superorganism that is more than the sum of its parts. A metaphor for the coexistence of humans with nature and with themselves?

Symbiosis is everywhere, not just in lichens. Symbiosis is the rule, not the exception. Even we humans are not individuals, as we live in symbiosis with microorganisms such as our gut biota. And we form part of a global ecosystem that can only survive if we learn to live in a sustainable symbiosis with our environment. Evolutionary biologist Scott Gilbert concludes that "we are all lichens," and Laurie Palmer's book *The Lichen Museum* uses lichens as a metaphor to understand human relations. Can we learn from lichens to live in harmony with ourselves and our environment?

→ An exquisite assemblage of delicate lichens growing together on limestone, featuring *Speerschneidera euploca* (above), *Endocarpon pseudosubnitescens* (middle below), and *Bagliettoa calciseda* (below to the right), conveying both diversity and individualism, concepts central to queer theory.

In John Wyndham's 1960 novel *Trouble with Lichen*, researchers discover a lichen that slows the aging process, the story building on the implications of this discovery to society, and on female empowerment. "Lichens are queer things," says a line in the book, more than two decades before the term "queer" was reclaimed as positively provocative by the LGBT(QI+) community and eventually became an umbrella term for non-normative sexual and gender identities.

In an essay written in 2015, sociologist David Griffiths developed a *Queer Theory for Lichens*, using the lichen example to oppose the traditional view of Darwinism, which claims that competition between individuals is the driving force of evolution. Symbiotic organisms challenge this view, subject to mechanisms of natural selection different from a traditional model of heterosexual reproductivity and vertical inheritance. Griffiths uses lichens to point out how symbiotic organisms deviate from the "primacy of sexual reproduction in biological and social discourses," and how the "symbiotic view of life can challenge this conservative and heteronormative approach to human and non-human sexuality and sociality." In symbiotic relationships, the association *per se* becomes the target of natural selection, acting upon the processes involved in mutualistic interaction.

Griffiths also refers to Lynn Margulis's discovery of endosymbiosis: certain organelles in eukaryotic cells, the mitochondria and (in plants and plant-like organisms) the chloroplasts, are of bacterial origin, having been incorporated in the course of evolution. Margulis argued that symbiogenesis, in a more general way, is the primary mechanism of evolutionary novelty, since effectively all organisms live in symbiosis. Griffiths emphasizes that multispecies relationships offer numerous alternative ways through which nature and life can progress and reproduce, challenging a "restricted and restricting view of human sexual reproduction." He suggests:

Thinking with lichens can potentially offer a queer way out of heteronormative narratives of human and non-human sexuality and sociality by decentering heterosexual biological reproduction as the only way that life (re)produces.

Lichens have also been described as fungi that discovered agriculture. In agriculture, farmers select particular genotypes of crops for their yield, tolerance of environmental conditions, or resistance against pathogens. The best-performing genotypes are then put on the market and shared with other farmers. Is this analogous to photobiont sharing

↑ Lichens are queer things. In John Wyndham's *Trouble with Lichen*, published in 1960, lichens are discovered to contain an anti-aging substance called "lichenin," and the novel explores the social consequences of non-aging and the empowering of women.

The two researchers in *Trouble with Lichen* do not reveal the identity of their "anti-aging" lichen. They describe it, however, as "grey-green leaves … variegated by yellow spots along the edges" and as "imperfect." The widespread *Crocodia aurata*, the Green Specklebelly, fits this description. Does it hold the secret to perpetual youth, or could it be adopted as the emblematic "queer" lichen?

in lichens? Many differences can be found. For one, lichen algae usually do not have sex while in symbiosis, while lichen fungi cannot have sex outside it. Indeed, they appear to depend on biochemical input from the alga to get the job done. If the same principle applied to the human–agriculture system, farmers would not be able to reproduce without some alarmingly direct involvement of their crops.

Apart from the question of who gets to have sex when and where, lichens transcend any normative views on biological reproduction. Besides varied approaches to sexual reproduction, lichens have evolved multiple ways to propagate their symbionts either separately or together, and the symbiotic partners even undergo horizontal gene transfer. Heteronormative sexuality, if it exists at all, plays a minor role in how lichens reproduce and pass on their traits to future generations.

So are lichens queer? And if we humans are all lichens, are we all queer? As Griffiths puts it:

We are all queer multispecies consortia, always already involved in countless and unpredictable constitutive relationships at all scales. The symbiotic view of life suggests that all organisms are involved in boundary crossings and gene-shuffling.

The word *normative* denotes conforming to a standard, which has also been equated with stereotyping. *Queer*, for its part, has been characterized as thinking and behaving "out of the box." The discovery of the lichen symbiosis more than 150 years ago can be considered one of the major early queer thought experiments in the biological sciences. It has become a symbol for novel scientific theories emerging against established norms. Our view of the lichen symbiosis is now once again entering a new era, with long-standing paradigms being called into question. Queer thinking marches on.

CETRARIA ISLANDICA

Icelandic Moss

Palate cleanser on the high tundra

SCIENTIFIC NAME	*Cetraria islandica* (L.) Ach.
PHYLUM, FAMILY	Ascomycota, Parmeliaceae
GROWTH FORM	Brownish ribbon-like thalli with characteristic whitish dots on the lower surface
SPECIES IN GENUS	15
HABITAT	On ground in heaths and tundra
NOTABLE FEATURES	Used for medicinal purposes in northern Eurasian cultures

Icelandic Moss is neither a moss nor strictly Icelandic. It is a squat, ground-dwelling lichen that grows both in the far north and in mountains at lower latitudes in dense mats of alpine tussock vegetation, heaths, and tundra. Even today it is one of the lichens that is most widely commercially available in a variety of products, including tea, body lotion, toothpaste, hair rinse, veterinary products, dyeing products, and, last but certainly not least, schnapps.

The pharmaceutical properties of Icelandic Moss were recorded in a medical book in 1671 and have been extensively studied over the centuries. The reference to Iceland in both its scientific and English names dates back to reports in the eighteenth and nineteenth centuries that it was a staple food of Icelanders, "without [which] they would … certainly perish," according to one account from 1874.

Extracts of the lichen have antibacterial properties, which have been attributed to the production of a fatty acid with the unwieldy name protolichesterinic acid. The lichen is still widely available in pharmacies across Europe, and its use in traditional medicine extends far beyond. The Yup'ik people of what is now called Alaska call it *aouk'*, and were recorded as chopping it up to add to various types of soups for flavoring. Traditional uses of lichens in Indigenous cultures tend to be under-recorded in the printed literature, and it is likely that people around the world have long known and valued this lichen.

Icelandic Moss is part of a group of species that are easily confused and were probably historically interchangeable in their usage, especially in the far north. The species all have similarly stiff, ribbon-shaped thallus lobes with eyelash-like projections along the lobe margins and cortex-free patches on the lobe undersides.

→ *Cetraria islandica*, showing its characteristic brown, ribbon-shaped lobes that have bare white patches on the lower surface.

Red Snow Tea Lichen

Versatile mountain-dweller of the Himalayas

SCIENTIFIC NAME	*Lethariella cashmeriana* Krog
PHYLUM, FAMILY	Ascomycota, Parmeliaceae
GROWTH FORM	Thallus shrubby, bright orange with the internal and basal portions blackish
SPECIES IN GENUS	About 10
HABITAT	On trees or sometimes on the ground
NOTABLE FEATURES	The branches feature a delicately ridged surface

When googling a particular lichen, one is usually presented with information on its taxonomy and natural history—which is sometimes plentiful and sometimes sparse. Not so the Red Snow Tea Lichen. In this case, Google hits look like a sales pitch, offering lichen material in sizable quantities from several suppliers. These represent various species of *Lethariella*, used to produce the famous red snow tea or as a herbal medicine.

Lethariella includes shrubby to beard-shaped lichens characterized by their orange color, caused by the pigment canarione. They appear similar to lichens of the genus *Teloschistes* (page 7), but are not closely related. Beard-shaped species are also reminiscent of dodders, parasitic plants in the genus *Cuscuta*.

Material sold to make red snow tea mostly represents the shrubby *Lethariella cashmeriana*, although it is often misidentified as the rare *L. cladonioides*. *Lethariella cashmeriana*

is found in the Himalayas, from western China to Pakistan, growing at high altitudes between 10,000 and 16,000 ft (3,000–5,000 m), and is usually epiphytic on coniferous or *Rhododendron* trees and shrubs.

The color of red snow tea is due to the orange pigment turning red in hot water. Apart from its traditional use in tea, studies of its biochemical properties suggest that infusions or extracts from *Lethariella* lichens may be effective against infections, high blood pressure, neurological issues, and even obesity and cancer. Indigenous people in the mountains of Pakistan, Tibet, Nepal, and Bhutan use a glowing, smoking mixture of *Lethariella* thalli and *Juniperus* branches as a hallucinogen. However, serious concerns have been raised about the sustainability of its harvest, given the demand for this slow-growing lichen.

→ A medicinal at risk from over-harvesting? The biochemistry of *Lethariella cashmeriana* (and related species) is well studied, and encompasses a diversity of substances, but its potential effects are likely caused by the orange pigment canarione.

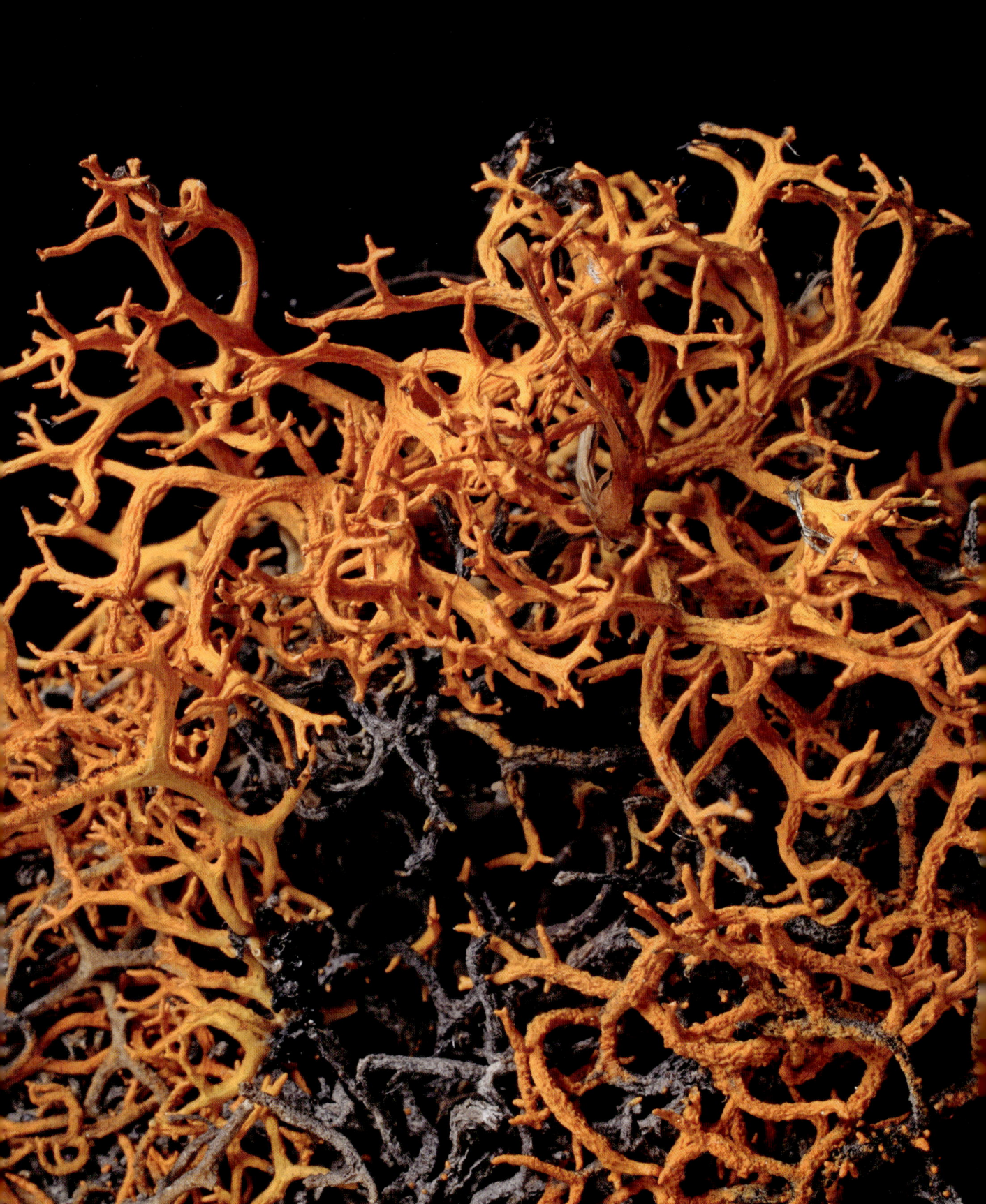

CIRCINARIA JUSSUFFII

Yousseff's Manna Lichen

The biblical manna reloaded

SCIENTIFIC NAME	*Circinaria jussuffii* (Link) Sohrabi
PHYLUM, FAMILY	Ascomycota, Megasporaceae
GROWTH FORM	Small pebbles resembling grain, loose on the ground
SPECIES IN GENUS	25
HABITAT	Arid regions in northern Africa to western Eurasia
NOTABLE FEATURES	Looks like compacted soil grains, moved around by wind

Various explanations have been floated for the biblical manna. The word "manna" has been used for various sugary plant exudates, and the term "mannitol" (for a sugar alcohol) derives from it. Potential sources include fungi, insect cocoons, and hardened honeydew produced by the Manna Scale insect, *Trabutina mannipara*, which sucks the sap of tamarisk trees, *Tamarix*. While all these match aspects of the biblical description of manna, none offers an explanation of how it could have fallen from the sky. But, given a sufficiently alarming weather forecast, lichens might do just that.

North African, Middle Eastern, and Central Asian arid regions harbor lichens of the genus *Circinaria*, forming loose thalli resembling seeds or grain, unattached to the ground and easily moved around by the wind ("vagrant"). These lichens were historically part of the diet of humans living in these regions, hence the name *Circinaria esculenta* (Latin *esculentus*, edible) for the best-known species.

There are accounts of these lichens "falling from the sky" after storms, in the region historically known as Mesopotamia, ranging from southeast Turkey through Iraq and Iran to Kuwait. The Kurds called these lichens "heavenly bread."

Work by the Iranian lichenologist Mohammad Sohrabi has shown that manna lichens represent different species within the genus *Circinaria*. Curiously, none of these are known from Egypt, where the biblical Exodus story is said to have taken place. The only species potentially occurring there is *Circinaria jussuffii*, known to the west, from North Africa, and to the east, from Iraq and Iran.

→ *Circinaria jussuffii*, the true manna lichen? Admittedly, one would have a hard time telling these lichens apart from compacted soil grains, let alone figuring out that they are potentially edible. The tiny grains are mostly only about $3/16$ in (5 mm) in diameter.

BRYORIA FREMONTII

Wila

First Nations winter staple food turned
molecular biology model system

SCIENTIFIC NAME	*Bryoria fremontii* (Tuck.) Brodo & D. Hawksw.
PHYLUM, FAMILY	Ascomycota, Parmeliaceae
GROWTH FORM	Fruticose, forming long, hair-like thallus filaments
SPECIES IN GENUS	50
HABITAT	Boreal forests, typically hanging from branches of conifers
NOTABLE FEATURES	Black, glossy perithecia on a whitish to brownish "thallus"

Cool conifer forests at northern latitudes are commonly festooned with enormous quantities of so-called hair lichens, involving members of *Bryoria* and *Alectoria*. Their thallus consists of hair-like filaments ranging from a few inches to over 3 ft (90 cm) long, and anything from sooty gray to brown to greenish-yellow in color.

Hair lichens are favored forage of deer (*Odocoileus* spp.) and Elk (*Cervus elaphus canadensis*), which often gather around fallen trees after storms to strip them of the lichens, and they are the sole winter food of Canada's endangered Mountain Caribou (page 205). Their value as food has also not been overlooked by humans.

Wila is the now widely used Secwepemc name for brown hair lichens that have been collected since time immemorial in the forest landscapes of inland western North America. Each of the many First Nations in this region has its own name for this valuable lichen, for good reason. Wila was, and in some areas still is, collected in late summer and formed into cakes that are baked in earth pits. It would be either consumed fresh after baking or saved for winter. Analyses of

the nutritional value of Wila suggest it is not a big energy hit, but nonetheless not negligible. In the past, it was probably an important means of survival in tough times when hunting was poor or after poor berry crops.

One of the most abundant Wila species is classified as *Bryoria fremontii*, and this species has become an important model for studying lichen symbiosis. Studies have shown that a toxic secondary metabolite, vulpinic acid, varies in its abundance in apparent correlation with the abundance of specific microbes, such as basidiomycete yeasts. The exact mechanisms behind this are still not understood, but *B. fremontii* offers an excellent system to study the interactions of fungus, alga, and other microbes because samples hang free in the air, comparatively untouched by contaminated soil or other lichens, and thalli can readily be extracted and characterized for genome analysis.

→ The lichen *Bryoria fremontii* can vary considerably in the content of the toxic yellowish secondary metabolite vulpinic acid. The yellowish forms, which are rich in this substance, were formerly recognized as a distinct species, *B. tortuosa*, but the production of vulpinic acid in the lichen appears to be connected to both habitat and shifting populations of yeasts in the lichen, not the species of fungus.

Highlighter Lichen

Brightly colored shrub lichens with a sinister backstory

SCIENTIFIC NAME	*Letharia vulpina* (L.) Hue
PHYLUM, FAMILY	Ascomycota, Parmeliaceae
GROWTH FORM	Chartreuse fruticose thalli that branch in irregular patterns, covered in isidia
SPECIES IN GENUS	6
HABITAT	On bark and wood of trees in mountain regions
NOTABLE FEATURES	One of the few lichens to be weaponized

Although many lichens are highly enriched with secondary metabolites, few are known to be poisonous. Perhaps the best-known poisonous lichen is the Highlighter Lichen, or Wolf Lichen, with as much as 5 percent of its dry weight made up of the toxic metabolite vulpinic acid. A mere ¹⁄₂₀₀ oz (150 mg) of the lichen can kill a mouse within 60 minutes; it is also a powerful antimicrobial.

The unmistakable chartreuse Highlighter Lichen is one of the suite of species that was given its scientific name by Linnaeus at the start of modern Western botany. He named it *Lichen vulpinus* (Latin *vulpes*, fox) because he had knowledge of its use by Norwegian hunters to kill foxes, as documented in an early work from the seventeenth century. Killing of Wolves (*Canis lupus*) using this lichen was first recorded after Linnaeus had already named it. Though not a happy thought, the name Wolf Lichen has stuck in several languages.

Over 200 years after Linnaeus, researchers discovered that the lichen long known as *Letharia vulpina* actually consisted of several species, which can be difficult to distinguish. They named one of these *Letharia lupina*, to finally give the wolf its due.

The culling of Wolves continues to this day (including in Linnaeus's native Sweden), and many people find it no laughing matter. The daughter of this book's second author sees a brighter message in this charismatic lichen, and has coined a new name: its hue almost perfectly matches that of a common modern office highlighter.

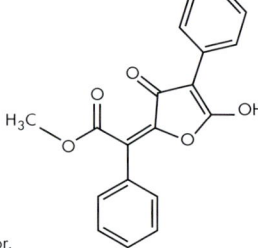

Powerful molecule

The pigment vulpinic acid gives this lichen its characteristic, chartreuse color.

→   *Letharia vulpina* is one of several species of highlighter lichens, forming a cryptic species complex. The two common species, with a thallus covered in tiny outgrowths (isidia) and hence appearing "rough," are *L. vulpina* and *L. lupina*. The latter has only recently been recognized as a separate species.

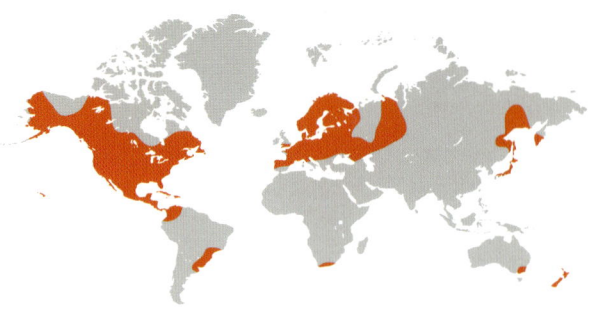

CLADONIA GRAYI

Gray's Pixie Cup

Revealing its secret under UV light

SCIENTIFIC NAME	*Cladonia grayi* G. Merr. ex Sandst.
PHYLUM, FAMILY	Ascomycota, Cladoniaceae
GROWTH FORM	Mid-sized to large, dimorphic, with erect, cup-shaped podetia
SPECIES IN GENUS	500
HABITAT	Temperate and tropical montane forests and woodlands
NOTABLE FEATURES	The cups produce numerous small granules for propagation

Ernst Haeckel was fascinated by the genus *Cladonia*. His famous lichen plate in *Kunstformen der Natur* contained no fewer than five representatives of this genus, three of them pixie cups of various shapes. With their cup- or trumpet-shaped erect thallus parts (podetia), pixie cups are among the best-known lichens, often depicted as the signature lichen *par excellence*.

The function of the cup-shaped podetia is not known. It may serve to disperse the tiny, granulose propagules lining the inner surface of the cups, the so-called soredia. One could imagine the cups filling with water during rain and then individual raindrops falling into these tiny pools and splashing out the soredia. However, soredia are also formed on the outside of the cups, and many other *Cladonia* species disperse quite successfully without forming cups, leaving the importance of this shape open to other interpretations.

With over 500 species, *Cladonia* is one of the most diverse genera of lichenized fungi. Its morphological diversity is extraordinary, ranging from squamulose to foliose thalli, to the typically dimorphic forms with erect podetia (including the pixie cups), to the shrub-like, cushion-shaped thalli encompassing the reindeer lichens. These diverse morphologies do not represent natural groups but evolved several times independently within the genus. *Cladonia grayi* is part of a complex of species in which chemotaxonomy, the distinction of species by secondary chemical compounds, plays an important role. The main compound in this species is grayanic acid, which gives it a light blue fluorescence under ultraviolet light.

→ *Cladonia grayi*, photographed under UV light to demonstrate the presence of grayanic acid.

GLOSSARY

apothecium (pl. **apothecia**) Disk-shaped fruiting body of an *ascomycete*, in which the surface of the *hymenium* is exposed.

ascomycete A fungus of the phylum **Ascomycota** (sac fungi), which forms sexual spores (**ascospores**) in a sac-shaped cell called an **ascus** (pl. **asci**); lichens formed by ascomycetes are called **ascolichens**.

basidiomycete A fungus of the phylum **Basidiomycota** (toadstools, bracket and jelly fungi, rust and smut fungi, and others), which forms sexual spores (**basidiospores**) at the tip of a cylindrical cell called a **basidium** (pl. **basidia**); lichens formed by basidiomycetes are called **basidiolichens**.

biofilm In a broad sense, a surface layer formed by an intricate community of small organisms, including bryophytes, algae, lichens, other fungi, and bacteria; in the strict sense, a thin layer of microorganisms sticking together.

biome A broad term for an ecosystem consisting of a range of different habitats, for example tundra or tropical rainforest.

cephalodium (pl. **cephalodia**) Part of a green-algal lichen *thallus* that contains cyanobacteria as additional *photobiont*; either internal or external, then often visible as small warts.

chloroplast An organelle within the cells of photosynthetic eukaryotic organisms (plants, diverse algae) in which photosynthesis takes place. In evolutionary terms, it is derived from a bacterial ancestor that was incorporated into the eukaryotic cell.

conidium (pl. **conidia**) Asexual spore produced by a fungus, usually at the tip of specialized *hyphae*.

cortex A zone of differentiated cells and secreted extracellular material delimiting the upper and/or lower layer of a lichen *thallus*, separating the interior parts from the environment or the substrate.

cyanobacterium A type of bacterium that is capable of photosynthesis and displays a blue-green color due to the presence of certain photosynthetic pigments.

endolithic Living beneath the rock surface.

gonidium (pl. **gonidia**) A historical term for the algae and cyanobacteria forming part of the lichen *thallus*, before it was realized what these cells represented.

heterotroph An organism that obtains its energy and nutrients from organic matter of other organisms and not from inorganic sources, such as light and minerals.

hymenium The layer in the fruiting bodies of fungi that contains the spore-bearing cells (for example the *asci* in *ascomycetes*) and additionally often densely arranged sterile *hyphae*.

hypha (pl. **hyphae**) Filament formed by end-to-end growth of fungal cells.

isidium (pl. **isidia**) Propagule(s) formed as cylindrical outgrowth of the upper *cortex* of the lichen *thallus*, also containing *photobiont* cells.

macrolichen Lichen with a three-dimensional thallus that can be easily separated from the substrate without destroying it (e.g., foliose, fruticose, and dimorphic lichens). In contrast, a **microlichen** produces a flat, more or less two-dimensional thallus and cannot be separated from the substrate without destroying it (i.e., crustose and squamulose lichens). Contrary to common belief, macro- and microlichens are not defined by size, as some foliose and fruticose macrolichens can be quite small (e.g., *Candelaria* or *Physcia*), while some crustose microlichens can reach sizes of more than 40 in (1 m) across.

medulla An inner layer of the lichen *thallus*, situated below the *photobiont* layer and composed of loose *hyphae* that are resistant to wetting, leaving large air spaces.

mitochondrion (pl. **mitochondria**) An organelle within the cells of most eukaryotic organisms. In evolutionary terms, it is derived from a bacterial ancestor that was incorporated into the eukaryotic cell by means of endosymbiosis.

mycelium The vegetative body of fungi formed by interwoven or conglutinated hyphae.

mycobiont The main fungal partner of the lichen symbiosis.

mycorrhiza Another form of fungal symbiosis in which the mycelium is closely associated with plant roots or even grows within the root cells.

perithecium (pl. **perithecia**) Ball- or pear-shaped fruiting body of an *ascomycete*, in which the surface of the *hymenium* is covered and opens only via a narrow pore through which the spores are ejected.

photobiont The algal and/or the cyanobacterial partner(s) of the lichen symbiosis.

photobiont layer The layer, usually below the *cortex* of a lichen *thallus*, that contains the *photobiont* cells, mixed with fungal *hyphae* attached to them.

photomorph/photosymbiodeme In lichens where the same fungus (*mycobiont*) associates with either green algae or cyanobacteria in different *thalli*, the different thalli are called photomorphs, as they usually differ in their morphology, including shape and color. The entire set of photomorphs is called a **photosymbiodeme**.

phyllosphere A term to describe the miniature communities growing on living plant leaves, especially in tropical rainforests, comprising lichens, bryophytes, non-lichen fungi, algae, cyanobacteria, other bacteria, and diverse invertebrates.

podetium (pl. **podetia**) The erect part of lichens that are composed of a basal, crustose, or squamulose thallus from which vertical structures emerge, usually to raise the reproductive structures above the ground.

polyol Sugar alcohol.

propagule Any structure produced by the lichen that serves for dispersal.

pycnidium (pl. **pycnidia**) A generally flask-shaped structure in which lichen fungi (and other fungi) produce asexual spores (*conidia*).

saprotroph/saprotrophic Fungi that use dead organic matter as a carbon source.

soredium (pl. **soredia**) Propagule(s) formed as a cylindrical outgrowth of the *photobiont* layer and *medulla* of the lichen *thallus*, composed of loose fungal *hyphae* wrapped around *photobiont* cells.

thallus (pl. **thalli**) Main body structure of a lichen; also used for leafless liverworts.

trichogyne A specialized, hair-like fungal *hypha* that receives fertilizing conidia during sexual reproduction and leads to the development of *apothecia* or *perithecia*.

zoospore General term for a spore with flagella that allow it to move.

USEFUL RESOURCES

Recommended books about lichen science and history, and guides to the identification of lichens.

Ahmadjian, V. 1993. *The Lichen Symbiosis*. John Wiley & Sons, Chichester: 250 pp.

Allen, J.L. and Lendemer, J.C. 2021. *Urban Lichens: A Field Guide for Northeastern North America*. Yale University Press, New Haven: 158 pp.

Andreev, M.P. and Himelbrant, D.E. (eds.) 2014–2023. *Флора лишайников России*. (*The Lichen Flora of Russia*). Volumes 1, 2, and 3. KMK Scientific Press, Moscow and St. Petersburg.

Brodo, I.M. 2016. *Keys to Lichens of North America*, revised and expanded edition. Yale University Press, New Haven and London: 427 pp.

Brodo, I.M., Sharnoff, S.D., and Sharnoff, S. 2001. *Lichens of North America*. Yale University Press, New Haven: 795 pp.

De Almeida, R., Lücking, R., Vasco-Palacios, A., Gaya, E., and Diazgranados, M. 2022. *Catalogue of the Fungi of Colombia*. Royal Botanic Gardens, Kew: 510 pp.

Dobson, F.S. 2014. *A Field Key to Common Churchyard Lichens*, 2nd edition. Privately published: 50 pp.

Dobson, F. 2018. *Lichens: An Illustrated Guide to the British and Irish Species*, 7th edition. Richmond Publishing, Slough, and British Lichen Society, London: 496 pp.

Fleig, M. and Grüninger, W. 2008. *Líquens: Flechten – Lichens*. EDIPUCRS, Porto Alegre, and Universität Tübingen: 217 pp.

Galloway, D.J. 2007. *Flora of New Zealand Lichens*, revised 2nd edition. Volumes 1 and 2. Manaaki Whenua Press, Lincoln: 2,261 pp.

Galun, M. (ed.) 1988. *CRC Handbook of Lichenology*. Volumes I, II, and III. CRC Press, Boca Raton: 297 pp., 181 pp., 147 pp.

Geiser, L., and McCune, Bruce. 2009. *Macrolichens of the Pacific Northwest*, 2nd edition, revised and expanded. Oregon State University Press, Corvallis: 448 pp.

Henssen, A. and Jahns, H.M. 1974. *Lichenes. Eine Einführung in die Flechtenkunde*. Thieme, Stuttgart: 467 pp.

Kashiwadani, H., Ohmura, Y., and Moon, K.H. 2020. 里山の地衣類ハンドブック (*Satoyama Lichen Handbook*). Bun-Ichi Sogo Publishing, Tokyo: 136 pp.

Kirk, P., Cannon, P., Minter, D.W., and Stalpers, J.A. (eds.) 2008. *Dictionary of the Fungi*, 10th edition. CABI Publishing, Wallingford: 771 pp.

Kirschbaum, U. and Wirth, W. 1995. *Flechten erkennen – Luftgüte bestimmen*. Ulmer, Stuttgart: 128 pp.

Marcelli, M.P. and Seaward, M.R.D. 1998. *Lichenology in Latin America. History, Current Knowledge and Application*. Companhia de Tecnologia de Saneamento Ambiental, São Paulo: 179 pp.

McCarthy, P.M. 1992, 1994, 2001, 2009. *Flora of Australia. Volumes 54 (Lichens – Introduction, Lecanorales 1); 55 (Lichens – Lecanorales 2, Parmeliaceae); 58A (Lichens 3), 57 (Lichens 5).* Australian Biological Resources Study and CSIRO Publishing, Canberra: 349 pp., 360 pp., 687 pp., 242 pp.

McCune, B. 2017. *Microlichens of the Pacific Northwest. Volume 1: Key to the Genera. Volume 2: Key to the Species.* Wild Blueberry Media, Corvallis: 218 pp., 755 pp.

McMullin, R.T. 2022. *The Secret World of Lichens: A Young Naturalist's Guide.* Firefly Books, Richmond Hill: 48 pp.

McMullin, R.T. 2023. *Lichens: The Macrolichens of Ontario and the Great Lakes Region of the United States.* Firefly Books, Richmond Hill: 608 pp.

Nash, T.H. (ed.) 2008. *Lichen Biology,* 2nd edition. Cambridge University Press, Cambridge: 486 pp.

Nash, T.H., Ryan, B.D., Diederich, P., Gries, C., and Bungartz, F. (eds.) 2002, 2004, 2007. *Lichen Flora of the Greater Sonoran Desert Region.* Volumes 1, 2, and 3. Lichens Unlimited, Arizona State University, Tempe: 532 pp., 742 pp., 567 pp.

Nimis, P.L., Scheidegger, C., and Wolseley, P.A. (eds.) 2002. *Monitoring with Lichens – Monitoring Lichens* [NATO Science Series. IV. Earth and Environmental Sciences 7]. Kluwer Academic Publishers, Dordrecht: 408 pp.

Orange, A., James, P.W., and White, F.J. 2010. *Microchemical Methods for the Identification of Lichens,* 2nd edition with additions and corrections. British Lichen Society, London: 101 pp.

Palmer, L. 2023. *The Lichen Museum (Art After Nature).* University of Minnesota Press, Minneapolis: 175 pp.

Piepenbring, M. 2015. *Introduction to Mycology in the Tropics.* American Phytopathological Society, APS Publications, St. Paul: 366 pp.

Poelt, J. 1969. *Bestimmungsschlüssel europäischer Flechten.* Cramer, Lehre: 757 pp. [Together with: Poelt, J. and Vězda A. 1977. Ergänzungsheft I. *Bibliotheca Lichenologica* 9: 258 pp. Poelt, J. and Vězda A. 1981. Ergänzungsheft II. *Bibliotheca Lichenologica* 16: 390 pp.]

Purvis, O.W. 2010. *Lichens,* revised edition. Natural History Museum, London: 112 pp.

Scheidegger, C., Keller, C., and Stofer, S. 2023. *Flechten der Schweiz. Vielfalt, Biologie, Naturschutz.* Haupt, Bern: 591 pp.

Schumm, F. and Aptroot, A. 2019. *Images of the Lichen Genus Caloplaca.* Volumes 1, 2, 3, and 4. Books on Demand GmbH, Norderstedt: 688 pp., 688 pp., 692 pp., 688 pp.

Schumm, F. and Elix, J.A. 2015. *Atlas of Images of Thin Layer Chromatograms of Lichen Substances.* Books on Demand GmbH, Norderstedt: 584 pp.

Seaward, M.R.D. (ed.) 1977. *Lichen Ecology.* Academic Press, New York: 550 pp.

Sharnoff, S. 2014. *A Field Guide to California Lichens.* Yale University Press, New Haven: 424 pp.

Singh, K.P. and Sinha, G.P. 2010. *Indian Lichens: An Annotated Checklist.* Botanical Survey of India, Ministry of Environment and Forests, Kolkata: 572 pp.

Smith, C.W., Aptroot, A., Coppins, B.J., Fletcher, A., Gilbert, O.L., James, P.W., and Wolseley, P.A. (eds.) 2009. *The Lichens of Great Britain and Ireland.* British Lichen Society, London: 1,046 pp.

Stenroos, S., Velmala, S., Pykälä, J., and Ahti, T. (eds.) 2016. *Lichens of Finland.* Finnish Museum of Natural History LUOMUS, University of Helsinki, Helsinki: 894 pp.

Swinscow, T.D.V. and Krog, H. 1988. *Macrolichens of East Africa.* British Museum (Natural History), London: 390 pp.

Van Herk, K. and Aptroot. A. 2004. *Korstmossen.* KNNV Uitgeverij, Soest: 423 pp.

Wei, Jiangchun 2020. 中国地衣型真菌综览(英文版)(精). *The Enumeration of Lichenized Fungi in China.* China Forestry Publishing House, Beijing: 606 pp.

Wirth, V. 2010. *Lichens of the Namib Desert.* Hess, Göttingen and Windhoek: 96 pp.

Wirth, V., Hauck, M., and Schultz, M. 2013. *Die Flechten Deutschlands.* Volumes 1 and 2. Ulmer, Stuttgart: 1,244 pp.

Zonca, V. 2022. *Lichens: Toward a Minimal Resistance.* Polity Press, Cambridge: 250 pp.

ORGANIZATIONS AND WEBSITES DEDICATED TO THE STUDY AND CONSERVATION OF LICHENS

American Bryological and Lichenological Society
abls.org

Australian Lichens
www.anbg.gov.au/lichen/

British Lichen Society
britishlichensociety.org.uk

Bryologisch-Lichenologische Arbeitsgemeinschaft für Mitteleuropa
blam-bl.de

California Lichen Society
californialichens.org

Consortium of Lichen Herbaria
lichenportal.org/portal

Global IUCN Red-Lists
lichenportal.org/portal/projects/index.php?pid=556

Grupo Brasileiro de Liquenólogos GBL
brazilianlichens.wixsite.com/website

Grupo Colombiano de Liquenología GCOL
licbiologia.udistrital.edu.co:8080/grupo-colombiano-de-liquenologia

Grupo Ecuatoriano de Liquenología GEL
grupoecuatorianodeliquenologia.blogspot.com

Grupo Latinoamericano de Liquenólogos
facebook.com/groups/glal1

Index Fungorum
indexfungorum.org

Indian Lichenological Society
indianlichenology.com

International Association for Lichenology
ial-lichenology.org

ITALIC
Italic.units.it

IUCN SSC Lichen Specialist Group
iucn.org/our-union/commissions/group/iucn-ssc-lichen-specialist-group

LIAS Glossary
glossary.lias.net/wiki/Main_Page

Lichen Determination Keys
archive.bgbm.org/BGBM/STAFF/Wiss/Sipman/keys/default.htm

Lichen Gallery
stridvall.se/lichens/gallery

Lichenicolous Fungi
lichenicolous.net

Lichenological Society of Japan
eng.lichenjapan.jp

Lichens
thomasbarlow.me/Lichens2021/n-Nc5xSv

Lichens Connecting People
facebook.com/groups/150880938305901

Lichens Marins
lichensmaritimes.org

Lichens of Great Britain and Ireland
britishlichensociety.org.uk/identification/lgbi3

Lichens of Subtropical Florida
seaveyfieldguides.com/Lichens/index.html

MycoBank
mycobank.org

Nordic Lichen Society
nhm2.uio.no/lichens/nordiclichensociety

OPAL Air Survey
imperial.ac.uk/opal/surveys/airsurvey

Pictures of Tropical Lichens
tropicallichens.net

Recent Literature on Lichens
nhm2.uio.no/botanisk/lav/RLL/RLL.HTM

Sharnoff Lichens Home Page
sharnoffphotos.com/lichens/lichens_home_index.html

Smithsonian Science Now—What's A Lichen?
video.ibm.com/recorded/124609758

Sociedad Española de Liquenología
ucm.es/seliquen

Società Lichenologica Italiana
lichenologia.eu

Tim Wheeler Photography
timwheelerphotography.com

Ways of Enlichenment
waysofenlichenment.net

INDEX

PICTURE CREDITS

The authors and the publisher gratefully acknowledge the permission granted to reproduce copyright material in this book.

Robert Lücking: Cover, Back cover, p3, p4 all, p5 all, p6, p7, p18, p23 top, p37, p42 left, p46 bottom, p47 bottom, p48 left & right, p49 right, p51 top right & bottom right, p52, p63, p65, p74, p75, p79, p82 top & bottom, p84, p86 top left & right, p86 bottom left & right, p87, p91, p93 top & bottom (redrawn from https://doi.org/10.1038/387463a0), p99, p105, p107, p112 all, p114 all, p115 all, p117 top & bottom, p119, p122, p123 top & bottom, p125, p126 top left & right, p127 top left & right, bottom left, p128 all, p129 all, p130 top & bottom, p131 top & bottom, p132, p133 bottom, p134 top & bottom, p137 top left & right, bottom, p138 top, p139 bottom, p140, p141, p143, p147, p149, p151, p159 bottom, p164, p165, p167 top left & right, middle right, bottom left & right, p169 top, bottom left, p170 (*Acarospora, Lobaria, Placopsis, Protoparmeliopsis, Rhizocarpon, Sarcographa, Stictis, Umbilicaria, Xanthoria*), p171 (*Candelaria, Lempholemma, Normandina, Pyrenula*), p178, p183, p185, p187, p189, p195 top left & right, bottom, p196 top & bottom, p197 left & right, p201 top & bottom, p204 top & bottom, p207 bottom, p209, p210 middle above & below, p213 top, p218 top left & right, bottom left & right, p220, p221 top & bottom, p222 left, middle, right, p223 left & right, p227, p229, p231, p245 left & right, p249, p252 top & bottom, p253, p254, p263, p264, p265, p267, p269, p271, p277.

Toby Spribille: p16, p21 top left, p59 bottom, p120, p138 bottom, 139 top, p159 bottom, p194 top, p237, p275.

Other photographers: p2: Shelly Benson; p4 (*Ochrolechia*), p73, p170 (*Ochrolechia*): Masumi Palhof; p11: Mohammad Sohrabi; p13 right: Bob Klips; p20 top left, p25, p31, p96, p273: Jason Hollinger; p21 top right: Carlos J. Pasiche Lisboa; p23 middle: Beata Guzow-Krzemińska; p23 bottom: Martina Tesarova/Jana Nebesarova; p27: Jonáš Gruska; p29: James Bennett; p33: Matt Goff; p35: Troy McMullin; p41: Corentin Loron; p42 right: Colin Purrington; p44: Allison Knight; p49 left, p51 top left: Michael Plewka; p51 bottom left: Nekko No Shippo (Japan)/PhycoKey by Alan L. Baker; p57 top left, middle right, bottom: Paul Diederich; p57 top right: Michel Langeveld; p59 top: Curtis Bjork; p61: Eva Barreno; p67, p76, p77: David Diaz-Escandon; p69: Mark Powell; p81 top & bottom, p89 left, right, bottom left: Leopoldo Sancho; p83 all: Daniel McCarthy; p92, p133 top: Bibiana Moncada; p95 top & bottom, p97: Daniel Stanton; p101, p121 top, p198 top: Volkmar Wirth; p103: Tessa Brunette; p113, p173 top: Einar Timdal; p121 bottom: Nicola van Berkel; p126 bottom left: Zacarias Lepista; p126 bottom right: Laura O'Connor; p127 bottom right: Noah Siegel; p135 top: Alejandro Huereca; p135 bottom: Gary Fast; p145: Larry Halverson; p154: Rosmarie Honegger; p155 top & bottom: Dong Ren/Xinli Wei/Xiaoran Zuo; p156: Dr. Alex Liu (University of Cambridge, UK); p157: Nobu Tamura; p159 top: Mary Droser; p161: Jonas Damzen (ex collection Jonas Damzen #7509); p169 middle left & bottom right: Pedro Crous; p171 (*Trichoglossum*): Jozef Pavlík; p171 (*Sclerophora*): Leif Stridvall; p171 (*Symbiotaphrina*): Adam Polhorský; p175, p176: Paul Cannon; p177: Tomasz Wilk; p181: Alexander R. Schmidt (University of Göttingen); p193 bottom: Ryan Brook; p198 bottom: Theo Llewellyn; p199 top & bottom: Adriano Spielmann; p206: Ricardo Miranda; p207 top: Enrique Ramírez García; p210 top: Marcelo Marcelli; p211 top left: Andrea Bernecker; p211 top right: Lidia Ferraro; p211 bottom: David Weiller; p213 bottom: Eddie Petryshen; p214: Laurens Sparrius; p216: John Skinner; p217 top & bottom: Gail Jones McMartin; p219 top: Heino Lepp; p219 bottom: Frank Bungartz; p225: Heath McDonald/Butterfly Conservation; p233: Eric Peterson; p239: Colin Chapman; p244 top: Ingo Haas/Botanischer Garten und Botanisches Museum Berlin; p247 top left: Christoph Scheidegger; p248 top: Anders Tehler; p250 left & right, p251 top left: Emily Donavan; p251 top right: Noah Siegel/Alissa

Allen; p255: Elke Zippel; p257 top: Bryony Ella; p257 bottom: © Kathrin Schlegel; p258 top: Oscar Furbacken; p258 bottom left & right: Kozue Mihashi; p259 top, upper middle, lower middle: Eileen O'Shea; p259 bottom left, middle, right: Kristen Morsches; p260 bottom: Thomas Barlow; p261: Anna Voytsekhovich.

Alamy Stock Photo: p43 left: Georg Stelzner; p205: Thomas Marent/Minden Pictures; p208: Guenter Fischer; p118: Michele Cornelius; p47 top Nigel Cattlin.

Shutterstock: p194 bottom: Grigorii Pisotskii; p248 bottom: H. Ko; p247 top right: Sam Thomas A.; p203: Tatu Lahunen.

American University Library: p260 top [Reproduced with the kind permission of the Government of the British Antarctic Territory].

ESA: p89 middle, p193 top [© ESA WorldCover Project/Contains modified Copernicus Sentinel data (2021) processed by ESA WorldCover consortium].

Eurolab: p171 (*Penicillium*).

Experimental Phycology and Culture Collection of Algae (EPSAG): p46 top left [2103 (CC BY-SA 4.0 DEED)], p46 top right [21.87 (CC BY-SA 4.0 DEED)/Pavel Škaloud]

Mint of Finland: p260 middle right.

Publisher Michael Joseph: p264.

Stadtarchiv Schaffhausen: p20 bottom right [StadtASH D I.02.521.04/0376].

The Field Museum: p251 bottom [© The Field Museum, Cat No. 19572, Photographer Elana Dux].

Creative Commons/Public Domain: p10: Yale Medical Historical Library (Public Domain/PD-US); p14: Museu Nacional da Universidade Federal do Rio de Janeiro (UFRJ) Brazil (Public Domain/PD-US); p22 bottom right (Public Domain); p43 right: Hagen Graebner (CC BY-SA 3.0); p171 (*Exophiala*): Medmyco (CC BY-SA 4.0); p171 (*Trichophyton*): Medmyco (CC0 1.0); p111: Sage Ross (Public Domain); p247 bottom: City Foodsters (CC BY 2.0).

Bayerische Staatsbibliothek/Public Domain: p244 bottom [https://bildsuche.digitale-sammlungen.de/index.html?c=viewer &bandnummer=bsb11198051&pimage=1237&v=100&nav=&l=de].

Biodiversity Heritage Library/Public Domain [https://www.biodiversitylibrary.org/item/]: p12 right [155796#page/725/mode/1up]; p13 left [155796#page/631/mode/1up]; p17 [151576#page/759/mode/1up]; p19 [151573#page/49/mode/1up]; p20 top right [91544#page/203/mode/1up]; p22 top [138200#page/7/mode/1up]; p22 bottom left [138200#page/17/mode/1up]; p172 [235808#page/3/mode/1up]; p210 bottom [595192#page/160/mode/1up]; p242 left [29685#page/348/mode/1up].

Giornale Botanico Italiano/Public Domain PD-IT: p54 [https://doi.org/10.1080/11263504109439808].

Münchner Digitalisierungszentrum Digitale Bibliothek/Public Domain [https://www.digitale-sammlungen.de/en/view/]: p12 left [bsb10870975?page=845]; p242 right [bsb11200259?page=669].

Staatsbibliothek zu Berlin/Public Domain 1.0: p243 [http://resolver.staatsbibliothek-berlin.de/SBB000149FE00010230].

ACKNOWLEDGMENTS

This book would not have been possible without the knowledge accumulated in countless studies by the ever-growing community of lichenologists, mycologists, and botanists, but also historians, artists, and philosophers interested in lichens, including all those that came before us. Our thanks go especially to those colleagues who, with their insight and expertise, helped to shape this book by providing unique and valuable images (separately credited in *Picture credits*) or critical information, including Arseniy Belosokhov, David Díaz-Escandón, Alejandro Huereca, Abigail Meyer, Walter Obermayer, Leo Sancho, Daniel Stanton, Heather Stewart-Ahn, Gulnara Tagirdzhanova, Lisa Willis, Alissa Allen, André Aptroot, Othmar Breuss, Irwin Brodo, Ryan Brook, Michelle Brownlee, Frank Bungartz, Paul Cannon, Peter Crittenden, Manuela Dal Forno, Paul Diederich, Emily Donovan, Elana Dux, Christopher Ellis, Edit Farkas, Alan Fryday, Oscar Furbacken, Trevor Goward, Allan Green, Martin Grube, David Hawksworth, Jason Hollinger, Ulrich Kirschbaum, James Lawrey, Gail McMartin, Troy McMullin, Joel Mercado-Díaz, Kristen Morsches, Jurga Motiejūnaitė, Leena Myllys, Peter Nelson, Pier Luigi Nimis, Yoshihito Ohmura, Eileen O'Shea, Sergio Pérez-Ortega, Christian Printzen, David Richardson, William Sanders, Christoph Scheidegger, Felix Schumm, Laurens Sparrius, Karina Wilk, Volkmar Wirth, Rebecca Yahr, Carmen Allen, Lore Ament, Frances Anderson, André Arsenault, Katie Barnett, James Bennett, Shelly Benson, Miriam Böhm, Claire Buchanan, Adolf Češka, Nathan Chrismas, Philippe Clerc, Janie Collin, Andrew Cook, Hugo Counoy, Amanda Davey, William Denton, Kendra Driscoll, Richard Droker, Sergio Favero, Adam Flakus, Fabiola Fonseca, Jean Gagnon, Ester Gaya, Nathan Gerein, Alice Gerlach, Allan Green, Mike Guwak, Curtis Hansen, Juán Hidalgo, Rosmarie Honegger, Deborah Isocrono, Suzanne Joneson, Jamie Kelly, Ilse Kranner, Teriyuki Kubo, Hans-Walter Lack, Scott LaGreca, Yung Mi Lee, Esteve Llop, Stefano Loppi, Joanna Marques, Maja Maslać, Didier Masson, Bruce McCune, Paolo Modenesi, Caleb Morse, Patricia Moya, Antonia Musso, Rikke Reese Næsborg, Thomas Nash, Leslie Nellis, Matthew Nelsen, Nastassja Noell, Nicole Nöske, Gerald Osborn, Alfredo Passo, Jozef Pavlík, Michael Plewka, Robert Sasata, Ayhan Şenkardeşler, Noah Siegel, Harrie Sipman, Pavel Škaloud, Ulrik Søchting, Mohammad Sohrabi, Marc Stadler, Tracy Thai, Göran Thor, Henk Timmermann, Benoit Trembley, Andrei Tsurykau, Saara Velmala, Andrus Voitk, Dennis Waters, Lilith Weber, Xinli Wei, Silke Werth, and Laura Wilson. Special thanks go to Chelsea Lau of Edmonton, Alberta (Canada), for realizing some of the beautiful custom graphics.

Special thanks go to the talented team of UniPress, in particular David Price-Goodfellow, Hugh Brazier, Slav Todorov, Wayne Blades, John Woodcock, Jennifer Dixon, Kate Shanahan, Natalia Price-Cabrera, Nigel Browning, Elaine Willis, Jan McCann, and Jan Ross. Their guidance, insistence, and constructive dialog helped to bring this project to completion.

This book has been a long journey and a steep learning curve since it started out in summer 2021, made possible through initial contact with our esteemed colleague and friend, Thorsten Lumbsch. Its assembly and completion would not have been possible without the support of our families, friends, and colleagues. Robert especially acknowledges the company, patience, and help of his wife Bibiana Moncada, herself a dedicated lichenologist. His son Mauricio and his daughter Natasha, always curious about lichens, kept him questioning the apparently obvious. Toby expresses his deep gratitude to his wife Viktoria Wagner for her infinite patience, and his insatiably curious daughter Daria, who read and reviewed many pieces of text and even proposed one or two new English names on some of her many hikes with him in the Canadian wilderness, as well as to Kathi and Pom Collins of Trego, Montana (USA), who provided a lichenological home away from home on so many occasions, a place where many turns of phrase were born.